Andre

For~~mulations~~

Architecture, Mathematics, Culture

1 Models, Machines, Manuals, and Cabinets   9

2 A Machine Epistemology:
Encapsulated Knowledge and the Instrumentation
of Architecture   31

3 Theorems Made Flesh:
The Architectonics of Mathematical Maquettes   61

4 Alternate Dimensions:
Measuring Space from Building to Hyperbody   101

5 A Virtuality Atlas:
Stereoscopic Drawing and Geometric Dream Space   141

6 Illusion Engines:
Drawing at the Speed of Light   181

7 Labyrinths, Topology, and Meinong's Jungle:
Architecture's Impossible Objects   221

8 All You Need Is Cube:
The Political Economy of Grid Space   273

9 Geometry's Mass Media:
Broadcasting Technique in Architecture's Hyperbolic Era   307

10 Crystal Collectives:
Architecture's Chemical Subcultures   343

11 Of Dabblers and Virtuosos   387

    Notes   395
    Acknowledgments   419
    Index   421

for *Athena*

1 Models, Machines, Manuals, and Cabinets

1.1   The architect Quintus Teal examines models of his hypercubic house. Robert Heinlein,
"And He Built a Crooked House," *Astounding Science Fiction* (February 1941): 69.
Source: Schneeman (Charles E.) Papers, D-238, Special Collections, UC Davis Library.
Courtesy of the family of the artist.

In Robert Heinlein's 1941 short story "And He Built a Crooked House," the iconoclastic young designer Quintus Teal proposes an architecture elaborated from the logic of four-dimensional crystals. Seeing the remarkable design freedoms and spatial economies of advanced mathematics, he aspires to move beyond the geometric limitations of his contemporaries toward a new architecture:

fig. 1.1

> Why should we be held down by the frozen concepts of our ancestors? Any fool with a little smattering of descriptive geometry can design a house in the ordinary way. Is the static geometry of Euclid the only mathematics? Are we to completely disregard the Picard-Vessiot theory? How about modular systems?—to say nothing of the rich suggestions of stereochemistry. Isn't there a place in architecture for trans-formation, for homomorphology, for actional structures?[1]

Mined from advanced mathematics, the complexities and con-tradictions of this architecture multiply, leading to the story's surreal conclusion (to which we shall return). Fantastic as it may seem, this fiction was paralleled to a remarkable degree by ever-deeper architectural fascination with modern mathematics in mid-twentieth-century design culture. In fact, exactly the methods that Quintus Teal mentions—such as modular struc-tures and homomorphology—were taken up by architects who saw in advanced mathematics a portal to dizzying new spaces and practices, unerringly precise and resolutely contemporary.

This book probes the facts of form and the frontiers of cre-ative rationality. It confronts these conjoined issues through a close reading of mathematics as it was drawn, encoded, imagined, and interpreted by architects in the mid-twentieth century, on the eve of digitization. The Swiss architect, artist, and educator Max Bill neatly distilled the general mathematical optimism in his 1949 essay "The Mathematical Approach in Contemporary Art" (Die mathematische Denkweise in der Kunst unserer Zeit) when he announced, "I am convinced it is possible to evolve a new form of art in which the artist's work could be founded to quite a substantial degree on a mathematical line of approach to its content."[2] Inspired partly by the beguiling mathematical maquettes and scientific models of eccentric geometric objects,

Bill eschewed more classical associations of geometric proportion to exult in abstract and modern varieties of mathematics. He saw fields like differential and topological geometry as templates for new design tactics. His hoped-for synthesis of technical precision and visual creation was symptomatic of a new subculture of design that absorbed mathematical ideas and representations, modeling the entire project of visual research on quantitative methods and on the pattern of scientific epistemologies.

At the nexus between architecture and mathematics are specific formulations—systems of common or co-opted method, representation, and explanation—which over the last century have fueled a more mathematical species of architectural authorship. Formulations are combinations, tinctures, tonics, potions (chemical or alchemical); things consumed, internalized, metabolized, digested. At the same time, a formulation is an entirely abstract statement of principle, a theorem, a catechism. Evoking both relational and morphological connotations, a formulation may be something atomically irreducible, something pure, something that surpasses obsolescence. Formulations may be intentional configurations of symbols, and associations with mark-making—writing, diagramming, and drawing—are inescapable. The "mathematical writing" that philosopher of science Brian Rotman has described is one mode of this mark-making, indicative of a "signifying practice" in which one may "make the signs and think the thoughts of mathematics."[3] Equally relevant from an architectural perspective are the signifying practices of mathematical drawing and diagramming, visual tactics resonant with their architectural analogues. Mathematical formulations furnish designers a means to exceed subjectivity through universal and inviolate facts of form, and a medium to estrange the intuition to the point of fresh creation. The historical lineages of mathematized design and its signifying practices offer periodic tables and catalogs of formulations—epistemic, logical, formal, social—that function as lexicons of mathematical writing and sourcebooks of mathematical drawing for precise practices of creation.

Design's Knowledge Cultures

History offers an abundance of hackneyed odes to the mystic transcendence of number in architecture:

> Mathematics is the majestic structure conceived by man to grant him comprehension of the universe. It holds both the absolute and the infinite, the understandable and the forever elusive. It has walls before which one may pace up and down without result; sometimes there is a door: one opens it—enters—and is in another realm, the realm of the gods, the room which holds the key to the great systems. These doors are the doors of the miracles.[4]

Enthusiastic as they may be, these sentiments of Le Corbusier are characteristic of a rhetorical obscurantism around mathematics in modern design that rests on a brittle bric-a-brac foundation of natural forms and mystic associations—the golden ratio, the nautilus shell, tedious references to musical rhythm, or the endless parade of anthropometric proportions. Both historians and popularizers have too often reduced geometry in architecture to mere symmetry or pattern. Such lazy associations have debased conversations on mathematics in architecture to a hopelessly caricatured level.

More subversive is the tactical but oft-misinterpreted view of Colin Rowe's influential essay "The Mathematics of the Ideal Villa." The qualification of the "ideal" villa is not innocent, and mathematics itself is also implicated in those Platonic associations. Despite the pretext of geometric proportion as the reputed liaison between Le Corbusier's Villa at Garches and Palladio's Villa Foscari, what Rowe offers us instead is a comparative anatomy of the architectural elements—rotundas, *piani nobili*, structural systems—in which mathematics per se plays the role of interpretive pretext. Yet the modesty of Rowe's narrow argument may persuade the reader that arithmetic and plane geometry comprise the whole of mathematics relevant to architecture. Rowe never makes such a crude claim, and indeed seems to sense the looseness and particularity of both Palladio's and Le Corbusier's proportional applications.[5] Yet his substitution of the narrower category of geometric proportion for the wider field of mathematics can leave

13

the impression of false equivalence. The more nuanced argument is lost and instead a caricature of Rowe persists as an alibi for a technically impoverished notion of mathematics in architecture.

Prior considerations of mathematics and architecture have often neglected the surprising and sometimes rigorous architectural encounters with a breadth of calculational systems over the course of the last century, including in the more abstract varieties of mathematics such as matrix analysis, polyhedral geometry, and information theory. Architects like Anne Tyng, Steve Baer, William Huff, Peter Pearce, and Lionel March took these methods seriously, inventively retooling them for design. They deployed a staggering range of mathematical techniques, from the elementary to the fantastically arcane, each adopted with various degrees of reverence and opportunism. For these and other architects, mathematics in design took on instrumental and teleological undertones, being both a set of techniques and a point of view freighted with ideological associations. Mathematical design is a dauntingly vast territory, practically a discipline in its own right, highly variegated, political, and contested.

Mathematics also furnished a singular medium through which architects could engage the physical sciences more broadly. Through the twentieth century and into the twenty-first, mathematics persisted as a kind of ur-science which oscillated between the poles of naturalism and antinaturalism. The philosopher José Ortega y Gasset claimed that "technology is man's reaction upon nature or circumstance. It leads to the construction of a new nature, a supernature interposed between man and original nature."[6] For certain architects, mathematics was a conceptual intermediary for the rapprochement of natural sciences with creative method. In fact, mathematics has often been a lens through which sciences such as biology and chemistry were understood and ultimately operationalized by architects. The role of mathematics vis-à-vis architecture essentially situates design's entire position relative to the natural sciences.

A Mathematical Turn

In *The Projective Cast*, the architectural historian and theorist Robin Evans invokes the durability of classical drawing techniques despite the irrevocable changes of Belle Époque science: "What I want to emphasize is that those aspects of nineteenth-century geometry that captured the imagination of modern architects could only be present in architecture metaphorically. Architects could not use the fourth dimension or hyperbolic space in the same instrumental way in which they used triangles and projections but they could allude to them, and that is what they did."[7] For Evans, accepted practices of drawing circumscribed the experimental horizons of architecture, ensuring that design encountered mathematics only in the narrow cul-de-sac of constructed geometry. For the architect, modern mathematics was an enticing but inaccessible mirage.

By registering a much wider spectrum of conceptual and methodological interactions between design culture and mathematical thought, I argue that architects went well beyond the mere allusions to advanced mathematics to which Evans claimed they were confined. While metaphorical invocations of mathematics were indeed among the earliest type, by 1970 certain architects had gradually assimilated quantitative means to address a range of unseen hyperdimensional forces. In the case studies that follow, architects are not merely alluding to advanced mathematical techniques but using them and training other designers to do the same. Moreover, Evans's limited identification of nineteenth-century mathematics with hyperdimensionality neglects epochal methods of drawing and modeling from other nascent branches of mathematics like geodesy, topology, crystallography, graph theory, and others that were appropriated by architects as engines to represent both designed buildings and more abstract spatial relationships. In fact, by the mid-twentieth century, calculational methods of transcription and encoding with roots in nineteenth and turn-of-the-century mathematics began to provide compelling alternatives to the accepted representational and methodological canons of architecture. A mathematical turn was well under way.

Yet, for a discipline that places considerable weight on an almost familial continuity of apprenticeship, reconstructing this mathematical turn presents many challenges. With no recognized lineage of mathematical experimentalism and no continuous institutional project of design mathematics, the transmission of

15

mathematical knowledge has been halting, sporadic, and incomplete. The inescapable gaps, fissures, and discontinuities that mark the terrain demand a forensic piecing together of a mosaic of vignettes, fragment by fragment, to hint at a more complete map.

While precursors extend two centuries before, convergences between mathematics and architecture culminate in the period between 1940 and 1970, a moment that parallels the rise and decline of more general technoscientific optimism in Europe and the United States. World War II brought technologization at an unimaginable scale and speed, a transformation that demanded a more scientifically and mathematically equipped society. By 1970, the mediagenic futurism of the Osaka '70 World Exposition had fully digested these developments as an information culture underwritten by mathematics. The intervening decades were shaped by innovations in scientific communication and education (scientific popularizations, the mass media of magazines, New Math) and economic life (expanding prosperity, technocratic governments, containerization-enabled global trade), and by a general embrace of the scientific orientation that mathematics exemplified. Most importantly, this was a moment when the concepts and images of mathematics were more widely diffused and consumed than ever before.

This period coincided with a sharper definition of the discipline of architecture as well as many of the sciences, including mathematics and physics. Architecture's increasing professionalization demanded more rigorous methods of design. Architects searched for more powerful and systematic tools to discipline an expanding range of dimensions brought by a complex and interdependent society, and mathematics was an appealing toolkit. Moreover, the beautiful and enigmatic drawings and models of the exact sciences exerted a gravitational pull on designers convinced that these representations decoded space in novel ways. Images representing buildings and those representing mathematical entities began to inexorably blur.

Drawings that adopted and adapted the conventions of both architectural and mathematical representation held particular currency in the conversation between disciplines. Since Alberti's development of linear perspective, the history of geometrized drawing has encompassed both mathematical and painterly

16

components. Architecture's affiliation has typically been to the painterly, emphasizing the effects and embodied spatial qualities achievable through drawings. But the more rarely invoked lineage of mathematical drawing—including practices like triangulated mapping, stereographic drawing, crystallographic drafting, or topological diagramming—was not completely alien from design, and it developed into a critical seed of mathematized architecture. Mathematical drawing in architecture was a specific kind of visual instrumentation, on the one hand a way to marshal quantified information, and on the other a way to discipline sight. It tended to dislocate architecture from more classical modes of drawing, particularly monocular perspective, and supplied an array of new drawing practices in which methods of calculation were explicitly visible. This adaptation of scientific ways of seeing and drawing to design was part of what Jonathan Crary has called the "ongoing mutation of visuality"—the gradual technological abstraction and transformation of how designers see and imagine.[8]

This text ventures a catalog of the practices of mathematical drawing and calculation in design that preceded and anticipated digitization, as well as an account of the formal compendia which became a cultural currency between modern mathematicians and modern architects. In a time of intensifying heuristic interaction between architecture, science, and technology, the actors of architecture were not only social—individuals, collectives, subcultures, and institutions—but also technological—artifacts, machines, media, and computational systems. These technologies facilitated the encapsulation, transmission, and consumption of knowledge and also became the historical media of intercultural exchange between architecture and other disciplines, particularly the hard sciences like mathematics and physics. Through this exchange, architects adapted the shapes and surfaces as well as the values and epistemic ideals of mathematics. Vibrant new mathematical subcultures within architecture were underwritten by media which encapsulated scientific knowledge in a form consumable to designers. The resulting hybrid subcultures mined a bricolage of distinct technical histories, with lineages of characters, objects, and institutions that intersected with the political and authorial histories of both architecture and mathematics but had their own independent structure and cultural agenda.

This uncharted terrain invites an unencumbered and direct assessment of the material and visual cultures of architecture and mathematics. It also demands an appropriate critical structure in which to relate these cultures. Rather than proceeding sequentially, the book adopts a more iterative and sedimentary structure. Each chapter revisits a similar timeframe, but attending to new tools, practices, and representations. The text thus loops, overlaps, and reconnects with itself to stitch a rhizomatic lacework of affinities and relationships across time and disciplines.

More indirectly, the text locates and weighs reason in the acts and representations of mathematized architecture. Placed amidst a geography of particular images, drawings, models, catalogs, and machines, rationality is more acutely qualified and circumscribed. Conversely, by triangulating mathematical rationality through its various representations, the modes of intellection and imagination that are adjacent to it and that reach beyond it also come into sharper relief. Through a close reading of mathematical design, the roles of rationality and its alternatives take on tangible contours.

Model and Instrument

In the cultures of architecture and science, models are physical embodiments that act as both conceptual and formal archetypes. The historian of science Soraya de Chadarevian argues that the consideration of models of a specific science "allows us to combine a study of experimental practice with an analysis of the public presentation and image of that science."[9] Public models became an interface for how architects perceived the content of the represented sciences. For designers like Max Bill, mathematical models served as canonical products, sanctified versions of how mathematically generated objects should appear. Designers deliberately signaled alignment with specific scientific concerns in part through the scientific models that they chose to collect as exemplars and to emulate in their work. Max Bill collected topological models, Frei Otto collected soap films, and Anne Tyng collected crystals, each collection enshrining a constellation of formal desires and methodological tactics. Models served as relics from the distant lands of pure science and lampposts toward that ideal realm.

18

If the model was a static formal archetype, architects also encountered dynamic mechanical instruments that encapsulated mathematical knowledge. Such technologies—drawing, surveying, projecting, and seeing machines, for example—encapsulated the underlying rules of which physical models were individual specimens. Drawing and seeing machines became metarepresentations of parametric families of individual models. Multiplicity allowed machines to become auxiliaries that would not merely record but amplify and transmute design ideas. The appearance of mechanical devices that could encode technical knowledge heralded a new epoch for both architecture and mathematics. Intricate sets of calculational rules could be embedded in physical or electronic mechanisms, allowing arcane processes to be "black boxed" and effortlessly reproduced. While the cybernetician Norbert Wiener first popularized the notion of the *black box*, it was particularly well described by philosopher of science Bruno Latour: "The word black box is used by cyberneticians whenever a piece of machinery or a set of commands is too complex. In its place they draw a little box about which they need to know nothing but its input and output."[10] Historically, modes of automatic mechanical recording, drawing, and seeing were quintessential black boxes that could nevertheless be co-opted to permute and create endless series of entirely new forms and visions. These tools fused perceiving, recording, remembering, knowing, and drawing into interdependent and mechanically modulated relationships.

The notions of encapsulation and encoding that were so central to a mathematical design process refer to essentially the same concept, but each attends to a distinct aspect of it. Encapsulation calls attention to the process of packaging, containment, or sequestration of knowledge in a physical or conceptual vessel. Encapsulation localizes knowledge and makes it portable, at the same time freeing it from social, institutional, and contextual entanglements. It is a process that exactly defines inputs and outputs, as well as the terms of interaction with the contained process. Encapsulation is always encapsulation *within*, and a container—a machine, device, or other medium—is implicated in it.

Encoding, on the other hand, attends to what is encapsulated and to the language of syntactic rules which transcribe knowledge into a form that can be operative. Encoding implies a translation, often from an intuitively apprehensible form to a language that is more technically specific. The product of the encoding process is not a static recording of a specific entity but rather a representation of rules that specify an expansive family of related entities. In concert, encapsulation and encoding offer a process that can radically amplify and distribute mathematical technique.

Reasoned Catalogs of Form

Architects register cultural affinities in part by furnishing their physical workspaces and social conversations with specific artifacts. Amalgams of artifacts shape the creative horizon of the designer as a repository of ur-forms and desire-objects which inevitably inflect the acts of visualizing and making. For the scientifically inclined designer, these furnishing artifacts often reference, encode, or embed knowledge-based content and methods. They act as vessels for the transmission of scientific ideas or as lexicons of scientific forms. Models and instruments are the mementos and material remnants of the liaisons between architecture and science.

In an analogous way, collections of scientific models and instruments form specific relational organizations of objects, and thus particular ontologies and epistemologies. These organizations are displayed and socialized in exhibitions and collection cabinets, or in printed catalogs and manuals. Cabinets organize canons of objects, while manuals detail the use of these objects. To use the philosopher Jürgen Habermas's distinction, cabinets and manuals embody the difference between "know-that" and "know-how." The cabinet collects the "thats" (a collection of forms) and the manual outlines the "hows" (a collection of operations).

The cabinet organizes scientific models or instruments in families and offers them not as disciplinary relics but rather as archetypes in conversation with broader visual and material cultures. The interpretation of relationships among these artifacts is necessarily spatial, with proximity implying affinity or distinction. Models that share a shelf are ipso facto members of the same conceptual category. The cabinet convenes a pantheon of specific cultural priorities, and, in Johanna Malt's words, "attempts to encompass and order knowledge in a single case."[11]

The ordering logic of cabinets varied enormously since their appearance in the Ancien Régime, but two poles mark a spectrum, roughly separated by the year 1872. In that year, the mathematician Felix Klein proposed his Erlangen Program, a project of classifying disparate geometries based on their algebraic characteristics. The Erlangen Program outlined a rigorous approach to taxonomic classification which reverberated through the whole culture of mathematics. The cabinets before 1872 tended to be inclusively encyclopedic, assembling artifacts from architecture, astronomy, geology, medicine, and any number of other fields. They also tended to unselfconsciously sort objects in a free arrangement of visual affinity or deliberate contrast. As the sciences specialized, they also narrowed their collections to theoretically explainable work objects. Particularly as scientific model making became a more organized and deliberate commercial activity, collections and catalogs after 1872 tended to reflect the taxonomic varieties that conformed to theoretical frameworks like Klein's classification system. In crude terms, cabinets tended to mirror the scope and order of epistemic systems that underpinned them. As the historians John Elsner and Roger Cardinal argued, "collecting is classification lived, experienced in three dimensions."[12]

Printed visual atlases are cabinets in book form, mirroring their classificational impulses and disseminating their contents to a wider visual culture. Since the fifteenth century, scientific drawings were compiled into indexed collations that became authoritative visual guides to particular disciplines. Perhaps the best known are the biological folios of drawings or paintings, such as those of John James Audubon or Ernst Haeckel, that documented varieties of natural flora and fauna. But the exact sciences—chemistry, physics, and particularly mathematics—had their own faunal catalogs of mathematized drawings. Crystallographers like Paul Groth used the intricate methods of descriptive geometry to meticulously draw the molecular configurations of chemical and mineral lattices, some naturally occurring and some abstractly speculative. Physicists like Felix Auerbach delicately documented electromagnetic, sonic, and optical fields as if depicting the rarest tropical bird species. Mathematicians like Felix Klein gathered novel forms in archives of drawings and collections of plaster models, built using architectural techniques.

These various encapsulations served to make technical knowledge portable, accessible, and recombinant—not only to scientists but also to those attracted to scientific visual culture, including architects.

Designers consumed these catalogs as references and critical touchstones, and replicated them through an architectural filter in their own design-oriented manuals of geometry. Architects co-opted the visual forms, the nomenclature, and ultimately the underlying methodologies of these catalogs. In this way catalogs were foundational media for technical and conceptual communication as well as cultural registration. They articulated the aspirations, conventions, and representations of a discipline. They elevated exemplars and bounded a canon.

Beyond collecting objects, compendia collect linguistic descriptions. Ann Blair, a historian of premodern information management, notes that "reference books were designed to store and make accessible words and things (*verba et res*)."[13] Catalogs are de facto lexicons, correlating vocabulary with form and circulating scientific terms of art that were traceable to specific mathematical subdisciplines. The process of their consumption was also a process of acculturation and linguistic assimilation. As distinct scientific words resurface in architectural conversations, one may trace definite lexical connections to the original scientific disciplines. Catalogs, indices, and compendia become time-stamped logs that identify the critical first moments of linguistic interaction between disciplines.

The notion of an atlas, library, compendium, or lexicon recurs as an essential artifact of both the visual culture of science and scientifically inclined design. For the historian of science Peter Galison, the physical "working objects" in catalogs of science are intimately bound up with "visual culture where it most distinctly intersects objectivity."[14] And the terrain of objectivity in nineteenth-century science, when many of these catalogs originated, was vast. As he writes,

> Nowhere does debate over the classification of such objects come into focus so strikingly as in the spectacular literary genre of the scientific atlas. There are atlases of anatomy, atlases of wounds, atlases of cells, atlases of clouds, atlases

of elementary particles, atlases of heads, atlases of peoples, atlases of stars—in fact, there are atlases of almost any collection of objects studied in science. . . . But binding them together is the aim of representing the basic species of a field of inquiry, usually addressed to practitioners with the aim of codifying existing data and to serve as the basis for further research.[15]

These atlases acted not only as references for specialists but also as distilled visual guides to the uninitiated scientific outsider. The great moment of catalog production in science coincided with the development of major classification theorems in surface geometry, topology, and crystallography, and cataloging was a byproduct of a broader taxonomic impulse of nineteenth-century science exemplified by Klein's Erlangen Program. Beyond the atlases of natural objects that Galison considers, scientists compiled catalogs of *representations*—physical models, drawing techniques, and even procedural algorithms—which were not mere logs of physical measurements but compendia of the abstract conceptual diagrams of scientific process. Compendia of mathematical technique produced by scientists were imbibed by designers for their own disciplinary reference and communication.

In distinction to the cabinet or atlas, the manual proposes a more operative constellation of elements. A manual is a "how to" guide or instruction set that proposes a specific mode of practice, using particular models or equipment. In a manual, the appearance of an object or element is less decisive than its function and suitability to a task. Instead of the definite spatial arrangements presented in a cabinet, manuals present recipes of actions to achieve a calculated aim. A manual offers a template of technique, and with it, applied tools for design. Since the time of the sixteenth-century architect Philibert de l'Orme, technical manuals have been integral to the knowledge culture of architecture. As Massimo Scolari has observed, "With the decline of the figure of the artist-scientist, a category to which de l'Orme in many ways belonged, theoretical knowledge became progressively extracted from practice and gathered together into specialized treatises. For those less cultured members of the profession, the model substituted other representational methods that required a profound knowledge of geometry."[16] Specialized treatises and manuals also passed between disciplines, as scientific manuals were co-opted for architectural uses. Manuals of drawing instruments, catalogs of mathematical models, or atlases of stereochemical and crystal forms exemplify the nominally scientific codices that found ready use within architecture.

Between 1940 and 1970, the increasing professionalization of architecture engendered new types of technical manuals that dealt with geometry and mathematics in a rigorous but pragmatic manner. Standalone manuals of mathematical form by and for architects began to appear, while mass-media disciplinary journals like *Architectural Record* and *Progressive Architecture* conveyed practical training on the construction of complex geometry like hyperbolic shells. The venerable genre of the geometric manual was reinvigorated by burgeoning mathematical subcultures within architecture.

The models, instruments, catalogs, and manuals that assembled the proximate ontologies and epistemologies of the scientifically inclined architect were products of both architectural studios and scientific laboratories. The chapters that follow sketch various routes by which these artifacts and accoutrements of the mathematical enterprise migrated from laboratory to studio.

Digital Mathematics

Within architecture, mathematics originated the cultural role that is now inherited by digital computation. Beyond the natural sciences like biology, minerology, chemistry, or physics, an account of mathematics and design necessarily implicates the epistemic issues raised by digital technology today. The mathematized design procedures encapsulated in computer software extended the formal field of architecture past classical expressions and media. But in the last decade, the computer itself grew beyond an expedient of image- and form-making to become a staggering repository for design knowledge, including mathematical knowledge. As digital computers were inexorably integrated into the practice of architecture, they became not only instruments to enable idiosyncratic and individual experimental programs, but also media for recording and communicating encoded design methods. This capacity to organize knowledge creates a *Wunderkammer* of algorithms, a cabinet of form- and space-making tactics. In other words, the computer has evolved into an epistemic system, a library of encodings that make accessible what were once the most obscure methods of deep disciplinary expertise. It is also an unprecedented mixing chamber in which an infinite recombination and hybridization of epistemic fragments is possible. Each unique algorithm, in all its exotic specificity and plumage, becomes a discrete element and building block in a collective edifice of architectural intelligence.

As its role and possibilities expand, architects and designers have begun to engage the computer on its own terms—in algorithms, code, geometry, and ultimately mathematics. Emerging techniques of artificial intelligence, machine vision, or simulated life are grounded in foundational mathematical theorems. To mobilize the full potential of computation, architects must embrace a rigorous standard of technical sophistication. At the same time, the "black box" encapsulation of methodology in code has fostered a kind of virtuosity in ignorance, in which technique is severed from knowledge. A critical antidote to this ignorance is a vision of how digitization was conditioned by historical positions toward calculation and mathematics in design, prior to any consideration of computational hardware. This thread of calculational thinking constitutes an essential subplot in the story of twentieth-century modernism, as well as in the continually unfolding exchange between architecture, technology, and the representational imagination.

This book deliberately avoids the well-trod and pseudo-canonical paths of early design computation and turns its attention instead to the prior mathematical excursions of liminal actors and experimentalists that were, in some sense, even more prescient. Boldface names will appear on stage, but more as cameos than as stars. The sporadic but collective nature of mathematical experimentalism demands a more ecumenical approach that can encompass a wider view of architecture's multivalent relationship with mathematics.

The Thought Collectives of Design Mathematics

As an alternative to a more linear account of mathematics in design typical of a narrative history, I propose instead an account that operates through the idea of the *thought collective*. The thought collective was a concept developed by the sociologist of science Ludwik Fleck in the 1930s to describe a critical cluster of mutually reinforcing researchers elaborating a common technical culture.[17] The architectural thought collectives that drew on mathematical influences were nourished through public exhibitions, science-popularizing books, and direct conversation with mathematicians. Architects in turn fostered nascent subcultures among themselves through lectures, magazine articles, personal contacts, or informal discussions, rumors, and even folklore. At a larger scale, certain institutions—like the German design school Hochschule für Gestaltung Ulm or the Canadian collective Groupe de Recherche Topologie Structurale—adopted projects of scientific rationalism that promoted mathematically inclined work. Of course, just as frequently there were maverick lone practitioners who, through insight or delusion, invented their own unique and idiosyncratic methods to harness mathematical possibilities.

What follows is a sketch of historical scaffolding organized as a sequence of "biographies of method"—episodes that configure a group of interlocutors, projects, institutions, tools, and cultural catalysts around a particular calculational technique or mathematical representation. Drawing on Lorraine Daston's notion of biographies of scientific objects, biographies of method attend to the peculiar and specific "embeddedness" of technique within culture.[18] It follows that every biography of method is also a sketch of a particular type of thought collective as well as a specific type of designer or experimentalist who draws synthetically on both scientific and architectural ideas.

At an organizational and interpretive level, this book adopts tactics of the cabinet as an associative device for thinking through transdisciplinarity. In a direct sense, it collects sundry artifacts of material and visual culture from across mathematics, physics, biology, chemistry, and architecture in a common critical frame and attends to their formal and logical lineage. Cabinets are not undifferentiated lists of particulars, but neither are they strictly determined, dimensioned, and deduced. Instead, they delimit both a field of artifacts and ways of relating and prioritizing those artifacts. The method of the cabinet attends simultaneously to the specifics of objects and the organization of epistemologies. Here we will observe the interactions between architecture and mathematics with the same spirit of associative generosity.

The eleven chapters that compose this study mark a broad territory of concerns, ranging from drawing to code, communication to curation, constructed vision to optical illusion. What they share is a curiosity about the myriad ways in which mathematics—including particularly mathematical notions of modeling and drawing—was spliced into the creative practice of design. Chapter 2 considers an original mechanical encapsulation of mathematical knowledge, the encoding of curve construction in complex drawing machines. Chapter 3 traces the artifacts of mathematical visualization—geometric maquettes—on their path into the lexicon of design. Chapter 4 unravels geodesy and mensuration, particularly the related practices of surveyed triangulation and photogrammetry, as a means to virtualize landscape and building. Chapters 5 and 6 treat technologies of mathematized vision, stereoscopic drawing and light-projected motion studies, and their design implications. Varieties of formal and functional topology appear in chapter 7, while cubic matrices and their economic implications are the focus of chapter 8. The intimate connection between mass media, hyperbolic geometry, and mathematical technique is unpacked in chapter 9. Perhaps the most integrated and coherent subculture is represented in chapter 10, with the advent of crystallographic design. Finally, chapter 11 speculates on the future of mathematics in the cultural mutations of architecture, as well as the role of knowledge in those mutations.

Throughout this arc, what was perpetuated across time was less a coherent system of architectural calculation than an intuition and a conviction that with the precision of mathematics, design could radically multiply its capacities. Through mathematics, architecture might master the hidden forces shaping design, synchronize technology and nature, and amplify beyond measure the intricate aesthetics of scientific visuality. For the designer, mathematics was a skeleton key to the exact sciences, what Robert Rauschenberg called "a catalyst for the inevitable fusing of specializations."[19] For certain architects, that fusion was a propulsive fuel to reforge a new architecture and a new architect.

In this journey we may be drawn down unexpected paths, as the young Quintus Teal was when he mused on the possibilities of a hyperdimensional crystal house:

> Think of the infinite richness of articulation and relationship in four dimensions. What a house, what a house. . . . Time is a fourth dimension, but I'm thinking about a fourth spatial dimension, like length, breadth, and thickness. For economy of materials and convenience of arrangement you couldn't beat it. To say nothing of the saving of ground space—you could put an eight-room house on the land now occupied by a one-room house. Like a tesseract—" "What's a tesseract?" interjects his compatriot. "Didn't you go to school? A tesseract is a hypercube, a square figure with four dimensions to it, like a cube has three, and a square has two. Here, I'll show you."[20]

And he does. He builds his tesseract crystal house, with vertiginously disorienting effects. Doors and windows became apertures into a folded space-time, a self-referential Möbius strip: "They climbed a fourth flight of stairs, but . . . they found themselves, not on the roof, but standing in the ground floor room where they had entered the house. . . . He flung open the front door, stepped through, and found himself staring at his companions, down the length of the second floor lounge."[21] They even detect an intruder—who, in fact, is the architect himself, shifted through space and time. With each successive door and window, itself a facet of this four-dimensional crystal, bizarre new tableaus appear, until they crescendo into a dramatic existential abyss:

> Teal lifted the blind a few inches. He saw nothing, and raised it a little more—still nothing. Slowly he raised it until the window was fully exposed. They gazed out at—nothing. Nothing, nothing at all. What color is nothing? . . . What shape is it? Shape is an attribute of something. It had neither depth nor form. It had not even blackness. It was nothing.

... He stared at the lowered blind for a moment. "I think maybe we looked at a place where space isn't."[22]

On the surface, the dizzying potency of mathematical structure is this very promise of taming the infinite, unbounded abyss of space with discrete and relentlessly rational encodings. But perhaps the power of mathematics in design is rather its capacity to bring architecture up to the edge of rationality and to peek beyond it. It is the power to transubstantiate the excess of reason into absurd poetry, to behold the waking dreams of science and dare to draw them.

2 A Machine Epistemology: Encapsulated Knowledge and the Instrumentation of Architecture

fig. 2.1 Among the aristocracy of prerevolutionary Paris, collecting expansive *cabinets de curiosités* was a diversion of elite refinement.[1] The most excellent among these cabinets were encyclopedic yet highly idiosyncratic collections whose specimens commingled art, natural history, mechanics, optics, minerology, cryptozoological taxidermy, and architecture. By the early eighteenth century, there were dozens of substantial collections in Paris alone, providing venues for conversation, entertainment, and research.

These *Wunderkammern* organized kingdoms of objects with a geography of implied conceptual relationships suggested by their spatial proximity. Mineral, animal, and vegetable specimens drawn from the classical kingdoms of nature jostled against man-made mechanisms and models in an uneasy epistemic continuum. As Lorraine Daston has observed, "It was the rule rather than the exception for most collections to embrace both *artificialia* and *naturalia*."[2] By being placed among natural artifacts, scientific and technological elements were implied to be continuous with the mechanisms of the universe itself. The cabinet was an aesthetic selection as well as an ontology—a revision of the classical *scala naturalis* into new maps of nature, artifact, and instrumentation.

Among the most monumental of these cabinets was the collection of Joseph Bonnier de la Mosson (1702–1744) in his *hôtel particulier* at rue Saint-Dominique, in genteel Saint-Germain, Paris. A financier and heir to a considerable fortune, Bonnier de la Mosson was also a voracious and eclectic collector. His sizable collection, consisting of eight substantial cabinets and numbering over a thousand objects, included eclectic fossils, minerals, pharmacological reagents, intricate ornamental wood turnings, wax models of human anatomy, chemical apparati, sophisticated mechanisms, and intricate architectural models. Between 1739 and 1740, at Bonnier de la Mosson's commission, the architect Jean-Baptiste Courtonne (1711–1781) drafted meticulous drawings of the collection in precise and measured detail.[3] Complemented by a comprehensive catalog of the estate sale of the collection after Bonnier de la Mosson's untimely passing in 1744, Courtonne's drawings reveal clues into the intellectual mapping of the material world in the twilight of the Ancien Regime.

Of Bonnier de la Mosson's eight principal cabinets, the *cabinet de mécanique et de physique*, which housed his collection of designed objects, was particularly marvelous. While natural specimens often took pride of place in such cabinets, historian of science C. R. Hill reminds us that "collectors also sought finely-made instruments and ingenious mechanisms embodying the principles of Newtonian mechanics and metrical science."[4] This cabinet encompassed at least four sets of shelves, filled with not only remarkable proto-industrial machines, but also polyhedral mathematical models, prosthetic limbs, and architectural maquettes. Immediately adjacent were surveying and drawing instruments interchangeably useful for an architect or natural philosopher, surveyor or mathematician, stonemason or astronomer. These machines formed a de facto shared instrumental lexicon that bound these disciplines.

Much like Bonnier de la Mosson's cabinet, a constellation of design machines that emerged from the creative practices of particular post-Enlightenment designers could be examined to locate the reasons these machines were used and the effects they produced for the design culture surrounding them. These so-called philosophical instruments encoded certain common methods of precise action—measuring a distant object, drawing a conic section, delineating a complex curve—that were equally useful to mathematicians, astronomers, surveyors, and architects. Such machines were not the exclusive property of one discipline. Instead, they migrated and mutated across disciplinary boundaries, and in so doing diffused technique through disparate epistemic cultures. This mechanical lingua franca allowed the ready and rapid exchange of technique through a mechanized intermediary, unifying aesthetic, engineering, and manufacturing activities around common tools. As we will see, these instruments encoded a specific kind of knowledge and rendered it shareable and consumable by anyone who could operate them.

Drawing instruments comprise a distinct class of representational and material artifacts shared across cultures of architecture and mathematics that mark a zone of vital interaction. They signal how technique, cultural conventions, and ultimately knowledge migrate as they limn what sociologist Karin Knorr Cetina has described as epistemic cultures, "those amalgams of arrangements and mechanisms—bonded through affinity, necessity, and historical coincidence—which, in a given field, make up how we know

2.1 *The Mechanical and Physical Cabinet of Joseph Bonnier de la Mosson*, drafted by Jean-Baptiste Courtonne. To the right are geometric and polyhedral models, with architectural and mechanical models at center. Source: Bibliothèque de l'Institut National d'Histoire de l'Art, collections Jacques Doucet, Cabinet de Bonnier de la Mosson – Recueil de dessins, NUM OA 720 (1–8).

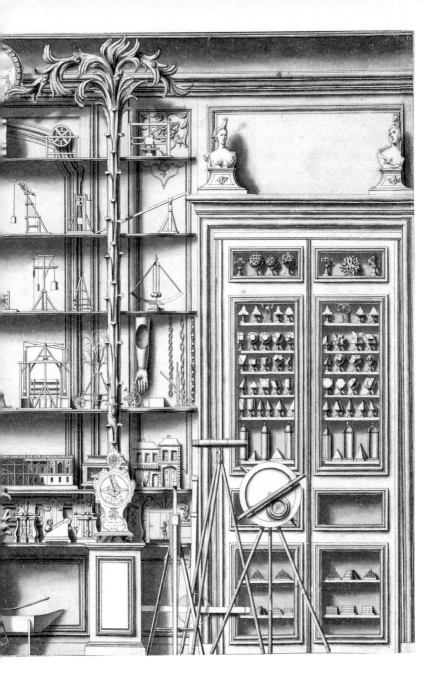

what we know. Epistemic cultures are cultures that create and warrant knowledge."[5] While Knorr Cetina intends these mechanisms broadly and sometimes analogically—the mechanisms of natural laws, theoretical deduction, or institutional policy—the physical machines of seeing, recording, measuring, and drawing are equally critical determinants of epistemic culture. Particularly for architecture, machines like sophisticated drawing instruments—and, by extension, digital drawing devices—play unique roles as transdisciplinary interfaces to mathematics itself.

As new machines like ellipsographs, conchoidographs, and planimeters began to encode the knowledge required to perform certain architectural drawing procedures, an epistemological distinction between design knowledge and instrumental knowledge emerged. On the one hand, we might define design knowledge as the architect's trained intuition of formal organization principles such as the relationship of parts to a whole, the ranges of material effects, and the appropriate use of geometry.[6] Instrumental knowledge is the narrower understanding of the procedures to successfully manipulate a certain type of technique or technology, which would include the ability to operate an instrument, a device, a machine, a process, or a software to intended effect.[7] In addition to its expedient use to accomplish a particular design, instrumental knowledge facilitates the creation of systems of interrelated technologies to systematically realize design agendas. The power of instrumental knowledge is its capacity to encode and encapsulate procedures in techniques and technology, and thereby make those procedures dramatically more accessible, communicable, repeatable, hackable, and transformable.[8]

Machines encapsulate operative aspects of design knowledge in an instrument's behavior. Geometric and spatial intuitions like "up and down," "inside and outside," or "center and periphery" are captured in drawing and surveying machines that encode specific mathematical forms. Encapsulation thus enables users to execute highly complex operations without explicit mastery of concepts underlying that complexity.[9] With instrumental knowledge of generative machines, the user can act as if she has the geometric knowledge it encapsulates.

There is a rich history of encoding and encapsulating geometric technique in sophisticated drawing machines used by architects to engage otherwise challenging design problems. Drawing machines became interfaces between architects and geometric knowledge systems, and placed architects in a new relationship with the act of calculation. The encapsulation of procedural technique in machinic tools transformed the application of mathematics in design from a process that required rigorous intellectual training to one that relied on instrumental intermediaries.

Mechanizing Curvature

The ambivalent role of technical knowledge in contemporary architecture is rooted in a debate on instrumentation in design dating at least to the fifteenth-century architects Filippo Brunelleschi (1377–1446) and Leon Battista Alberti (1404–1472). Their arguments illustrate the divergent positions on the role of machines in the genesis of architecture as discipline. Brunelleschi's singular architectural oeuvre encompassed not only the design of buildings but also the invention of mechanical instruments to construct those buildings: specialized hoists, jigs, and lifts. His dome for the cathedral of Santa Maria del Fiore in Florence (completed 1436) is the most surpassing manifestation of this integrated approach, as his burial inscription testifies: "How Filippo the architect excelled in invention is shown not only by the beautiful shell of his famous temple but also by the various machines that he invented with divine genius."[10] For Brunelleschi, architecture integrated the instrumental knowledge of devices and machines needed to holistically achieve the conceptual aims of design.

Leon Battista Alberti took a different position on the architect's role in his influential treatise *De re aedificatoria* (1452):

> Him I consider an architect who, by sure and wonderful reason and method, knows both how to devise through his own mind and energy, and to realize by construction, whatever can be most beautifully fitted out for the noble needs of man, by the movement of weights and the joining and massing of bodies. To do this he must have a knowledge of all the highest and noblest disciplines.[11]

Alberti's view of the architect and her training is so widely accepted as to be nearly canonical. While he definitively acknowledges the role of knowledge (or at least "knowledge of all the highest and noblest disciplines") in design, he yet qualifies the role of certain technical knowledge: "I should explain exactly whom I mean by an architect: for it is no carpenter that I would have you compare to the greatest exponents of other disciplines: the carpenter is but an instrument in the hands of the architect."[12] In this simple divorce of the architect from the carpenter, Alberti tacitly distinguishes between design knowledge and instrumental knowledge and implicitly calls into question the value of technical knowledge for the architect.

In a sense, Alberti is recapitulating a much older polemic originating with Vitruvius, for whom, as Massimo Scolari has observed,

> Machines and instruments were in the realm of the *mechanicus* rather than the *architectus* who directed their use. In alluding to certain machinations, Vitruvius's aim was to extol the *sophia* of the architect rather than the *téchne* of the mechanic, to show off the learning of those with knowledge of the principles, rather than the technical know-how of those who handled materials.[13]

Architectural historian Mario Carpo has shown that Alberti was profoundly interested in mechanics, particularly for the control and replication of representation.[14] Yet interest is not identical with expertise, and Alberti's machines lacked certain key innovations, such as screw-activated clutches and load positioners, that Brunelleschi had already introduced and embraced.[15] These lapses have led some critics to argue that Alberti's understanding of machine engineering was scarcely more that casual. For Alberti, it seems, instrumental knowledge of operative machines was not indispensable to design knowledge but a vestige of pretheoretical design practices. Brunelleschi's and Alberti's divergent approaches to the machine demonstrate the controversial role of instrumentality in architecture at least since the Renaissance.

Notwithstanding Alberti's alienation of architecture and
instrumentation, over the ensuing centuries the increasing formal

ambitions of architects demanded ever more sophisticated auxiliaries to design complex geometric forms such as double-curved vaults, delicate squinches, intersecting arches, or intricately twisting staircases. The need to supplement the architect's design knowledge with precise geometrical and mathematical procedures became especially pressing with the curved deformations and projections of the European baroque. Initially, the requisite geometric knowledge was developed by architects themselves. Robin Evans has written extensively of the advancements in architecture that enabled perspective and projection to be employed to design apparent and actual spatial distortions.[16] The work of the mid-sixteenth century French architect Philibert de l'Orme (1514–1570), who applied such techniques of distortion to the design of extraordinary vaults, was distinctly influential.[17] In this emerging lexicon of geometric figures, the ellipse was a particularly fertile area of architectural investigation. Procedurally, the ellipse is the specific conic section that results when a circle is projected onto a ramp or an oblique plane. During the baroque period, the ellipse took on its most recognizable role as a frequent trope of distorted and exaggerated ornament. Gradually, however, architects realized that the construction of an ellipse was also essential to the delineation of certain complex spaces. For example, the ellipse appears as the intersection of two cylinders of equal radii, and thus as the rib of a cross vault; or as the intersection of a single cylinder with an oblique wall or ceiling, and thus as an embrasure. This geometric understanding of the conic sections elevated the status of curved geometry from merely decorative to inherently spatial.

The drawing procedures promulgated by de l'Orme and others to represent curved forms were sophisticated and time consuming, and resembled intricate cartographic constructions. Without a means to automate drawing complex curves, these methods demanded the tedious sampling and projection of series of points in order to manually reconstruct complex shapes. Moreover, the complexity of these design operations required knowledgeable and astute practitioners. Some architects, engineers, and cartographers fluent in such methods turned toward toolmaking to encapsulate the knowledge of these geometric operations in specific drawing machines. In this instrumentalization, the toolmakers deployed various methods for the mechanical construction of conics developed by mathematicians such as René Descartes, Isaac Newton, and Colin Maclaurin, including techniques for the graphic calculation of ellipses, parabolas, and hyperbolas.[18] By the eighteenth century, advanced toolmaking, once merely an auxiliary discipline to support the drawing practices of architecture or astronomy, had become a sophisticated discipline in itself, with its own encyclopedic references and disciplinary projects. Nicolas Bion's *Traité de la construction et des principaux usages des instruments de mathématique* (1709) was a compendium of the state of the art of toolmaking during this period. Bion's work taxonomized the typical compasses and dividers as well as more ambitious implements such as trisectors and elliptical trammels, which were all applicable in architectural contexts. Treatises like Bion's confirmed the codification of procedural instruments as a locus of complex geometric knowledge.

As mathematical processes were mechanized, the machine became an intermediary that could tune, modulate, and permute geometry in strictly defined dimensions. In choosing to employ a mechanical drawing instrument, the user had to enact a defined sequence of operations, each one a distinct motion that altered the final figure in a narrow and determined way. With the adjustment of a screw, replacement of a gear, or movement of a hinge, the very kinesthetics of drawing were made less haptic and more discretely regimented. This type of manipulation was completely different from the more fluid and integrated gestures of hand drawing that it replaced. Instead of continuous interaction between hand and mind, drawing machines segmented, serialized, and organized the physical act of drawing into discrete categories of mechanical motion.

fig. 2.2 Eighteenth-century toolmakers proposed an ever-expanding panoply of methods for the construction of complex curves that exploited an Enlightenment understanding of mechanical motion. The Italian mathematician Giambattista Suardi's (1711–1767) work in particular is stunning for its range and advances in this domain. In his captivating book *Nuovi istromenti per la descrizione di diverse curve antiche e moderne e di molte altre* (1752), Suardi proposed two particularly unusual instruments: the first was a conchoidograph for the drawing of conchoid curves, and the second was a sophisticated "geometrical pen."[19] This "pen" was actually a compound machine designed by Suardi to mechanically draw a huge library of curves ranging from ellipses to complex epicycloids. The geometrical pen leveraged the combinatorial variety of curves that could be generated by the synchronized motion of two or more circles rolling within or around each other's circumferences. By swapping a series of simple gears on the pen, Suardi could vary the radii, tracing speed, and length of the drawing arm to produce hundreds of distinct curve types including straight lines, circles, ellipses, and many other higher-order compound cycloids.[20] The sundry categories of cycloid curve had a cosmological significance, as they were the curves that marked the planetary paths as they are observed from Earth. In architecture, they were the basis of ornamental patterns and certain types of column fluting.

Devices like the geometrical pen obviated the need for the user to possess the geometric design knowledge encapsulated in the instruments. The miraculous and effortless fecundity of Suardi's tool inspired many later devices that drew on its basic design. Ultimately the geometrical pen became a standard of sophistication and versatility by which later nineteenth-century drawing machines would be judged. As a consequence of the encapsulation of such complex generating processes in relatively simple devices, the stage was set for an even more profound mechanization of form to follow.

Drawing Machines, Knowledge Machines

New geometric methods created more urgent demands for tools that could rapidly automate complex drawing operations. Enterprising inventors were eager to meet that demand with a range of ingenious instruments that collapsed the labor and time of rigorous geometric constructions. For architects, the automation of conic sections—ellipses, parabolas, hyperbolas—was particularly helpful, since these curves frequently occurred as intersections and projections of cylinders and cones in vaults, windows, and embrasures. The English mechanical engineer John Farey Jr. (1719–1851) proposed one of the first influential mechanical tools for the construction of ellipses in 1813.[21] Known as an ellipsograph, it dramatically enhanced precision and control

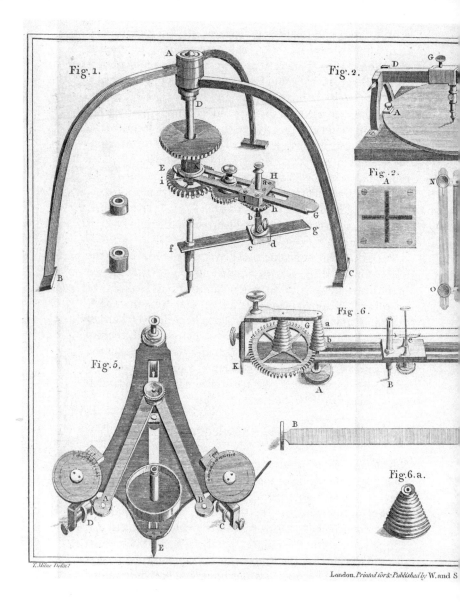

T. Milne Delin.^t

London. Printed for & Published by W. and S

2.2   A selection of drawing instruments, including Giambattista Suardi's geometrical pen, made by the English machinist George Adams, 1791.   Source: George Adams, *Geometrical and Graphical Essays: Containing a General Description of the Mathematical Instruments Used in Geometry, Civil and Military Surveying, Levelling, and Perspective* (London: W. and S. Jones, 1813), plate XI.

PLATE XI.

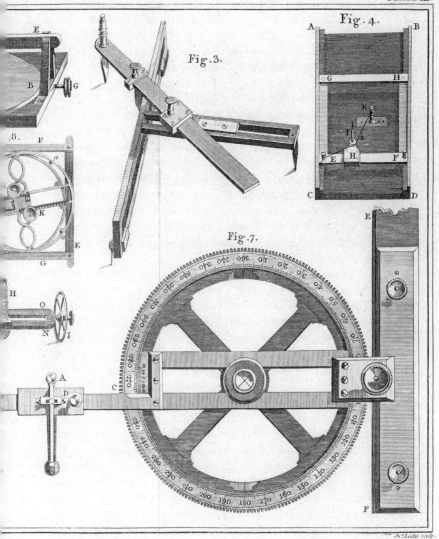

Fig. 3.

Fig. 4.

Fig. 7.

.8.

, Holborn, as the Act directs. June 1.1791.

Ino. Lodge sculp.

over previous methods of ellipse construction like the elliptical trammel. Soon tools began to emerge to draw other, more complex curves as well, often designed by architects for specifically architectural applications. Various types of helicographs for the drawing of logarithmic spirals, the volutes of Ionic columns, and the traces of spiral staircases were introduced and patented by the English architect Francis C. Penrose (1817–1903) in 1850.[22] Penrose had made meticulous studies of the various curvatures of Greek monuments between 1845 and 1846, and his later publication of these results explicitly related them to helical constructions.[23] In 1871, the English architect Edward Burslow designed a complex ellipsograph that was considered among the best by his contemporaries. English machinist William Ford Stanley's (1829–1909) conchoidograph drew the entasis of classical columns, following principles detailed by Suardi a century before.[24] Ellipsographs and their cognates thus opened new possibilities of exact geometric construction.

Patterned ornamentation presented a different application of mechanical drawing, one which expanded the definition of drawing itself. Instead of facilitating exact constructions of preconceived forms, combinatorial machines like Suardi's geometrical pen generated a menagerie of surprising and unexpected patterns and shapes. Among the products of these combinatorial engines were rococo figures of *guilloché*, a vertiginously intricate type of graphic tracery that was virtually impossible to execute by hand. *Guilloché* was originally an architectural term for a specific type of winding linework pattern typical of classical friezes, composed of braidlike chains of arcs.[25] The device that ultimately made *guilloché* infinitely reproducible was the geometric lathe. A kind of twin of Suardi's geometrical pen, hovering between a drawing machine and a fabrication tool, the geometric lathe was intended not for drawing but for engraving differentiated procedural patterns onto an underlying printmaking plate. By means of complex roulette curves such as hypotrochoids and epitrochoids, derived from the compound movements of its gears, delicate moiré-like patterns were traced on a substrate of metal or wood. This lathe was developed by the American machinist Asa Spencer in 1812 and was quickly applied to etch *guilloché* as an anti-counterfeiting measure in early United States banknotes. By

1819 the device was being adapted for use in the United Kingdom as well.[26] Mechanization made absurdly complex drawings a quotidian affair.

Among the most sophisticated combinatorial drawing devices was the clampylograph, designed by the physicist and meteorologist Marc Dechevrens (1845–1923) around 1900, and refined in 1903 by F. N. Massa.[27] This intricate device featured several compound motions but could replicate a vast range of curves, over 979 distinct types, including all of the conic sections and a range of cycloids, epicycloids, and hypotrochoids.[28] Its complexity ensured it was fantastically rare, and the Institut Poincaré in Paris has one of the very few known models. Nevertheless, the clampylograph probably represented the pinnacle of sophistication in the fin-de-siècle construction of curve-generating devices. With such machines, for the first time, drawing fantastically intricate figures was a matter of simple mechanical operation.

As drawing machines rigorously sequenced and structured the interactions of the user, they remixed the terms of engagement between architects and geometry. They narrowed the range of possible inputs, but they also dramatically expanded the speed and range of possible outputs, presenting marvelous new potentials in the creation of drawings. Combinatorial drawing machines mathematized the act of drawing in two ways. First, they shifted drawing from a purely visual exercise to an indexed process of calculation. The volute of an Ionic capital or the entasis of a Doric column could be encoded as lists of precisely reproducible numeric adjustments for the appropriate drawing device. The free mannerisms of classical orders were indexed on a continuum of tabulated form. Combinatorial machines produced seemingly disparate graphic patterns as manifestations of the same mathematical logic and mechanical motions, tuned in slightly different ways. Second, drawing machines shifted the burden of accuracy and fidelity from a fastidious human draftsman to a relentlessly reliable and invariably precise machine. What was once a feat of human attention and discipline was accomplished by the grinding motion of gears. The most exact sensitivities of the architectural eye were satisfied through calculation rather than intuition.

Manuals of Geometric Knowledge

Driven both by developments in the mathematics of curvature and by the emergence of more exact and scalable machine reproduction technologies, the encapsulation of geometric knowledge in machines accelerated in the nineteenth century. New technical developments catalyzed a reciprocally reinforcing cycle: the techniques of precise geometric drawing drove the development of particular instruments, while the availability of exact instruments facilitated new types of drawings and thus new geometric innovations. As machines used to describe geometry became more integral to the process of architectural drawing, the boundary between instrumental knowledge and design knowledge was inexorably blurred.

During the nineteenth century, a culture of drawing practice that spanned engineering and architecture fostered this virtuous cycle of geometric advancement. As the historian Andrew J. Butrica has noted, by 1840 admission to the most selective French schools across a range of disciplines rested on fluency in drafting and geometry.[29] This embrace of technical drawing as a common medium of representation followed the codification of descriptive geometry through the work of the French mathematician and physicist Gaspard Monge (1746–1818). Drawing on techniques honed in decades of military architecture, Monge's *Géométrie descriptive* (1799) was a general manual for the mathematized drawing and transmutation of diverse spatial forms from fortifications to furnaces, engines to embrasures. Descriptive geometry rested on mathematical manipulations of curves, surfaces, and their complex intersections as sequences of coordinated projections. It was an abstract but supremely practical science. Monge saw descriptive geometry as particularly relevant to the conception and operation of machines, and claimed that mastery of spatial calculation was key to France's industrial revolution.[30] Yet Monge's methods also redefined the geometric technique of architecture, arming designers with radical new tools for conceiving complex forms.

If descriptive geometry elevated the disciplinary knowledge of architecture and coordinated it with the visual practices of engineering, it also indirectly created demands for new kinds of instrumental knowledge. Monge's sequential projections were onerous to produce, and without the aid of mechanical tools, production of the requisite drawings was intricate and time-consuming work. Projective methods required extensive use of auxiliary views at irregular or oblique angles, which in turn demanded mechanical compasses. Drawing curved geometries was even more labor-intensive: point-by-point projections and interpolations were often essential for even the simplest of curves. Many of the distorted shapes derived through projective constructions were better drawn by machine tools, which were quicker and more accurate than point-by-point plotting. The complexity of design ambitions in the nineteenth century thus contributed to the need for instruments that could encapsulate these geometric operations in simple and repeatable ways.

If drawing machines sequenced and serialized the physical act of drawing, the allied technical manuals derived from Monge's work sequenced the intellectual acts of mathematical and geometrical manipulation. Geometric manuals were the concentrated distillations of disciplinary expertise. Though their ostensible aim was the diffusion of design knowledge, they also deconstructed and atomized previously organic drawing processes into discrete, serialized, and procedural instruction sets. In this sense they were a step toward the encoding of the same instruction sets in instruments, a beginning of the decoupling of technique from knowledge. In fact, sophisticated curve-drawing machines were only conceivable because their underlying processes been explicitly set out in such drawing manuals. In defining processes as deterministically as possible, manuals brought complex geometry ever closer to machine instructions.

During the nineteenth century, Monge's work encouraged the proliferation of technical manuals that applied his methods to all aspects of design, craft, and fabrication, particularly the stereotomic gymnastics of carpentry and stonecutting. These manuals sat somewhere between texts of disciplinary technique and catalogs of tectonic form, and they promoted a sumptuous formal range in the architecture of early nineteenth-century France. An early but impressive example was Jean Paul Douliot's *Traité spécial de coupe des pierres* (1825). Douliot was among the first to apply Monge's ruled surfaces, shapes swept by the quasi-mechanical movement of a generating line through space, to the creation of curved architecture. The result was a catalog of cylindrical, conic, ellipsoidal, and complex doubly curved ruled surfaces rationalized into stones of the appropriate geometry.[31] Douliot and his generation of manual makers rationalized speculative geometry into architecture.

Innumerable texts in the vein of Douliot followed. There were geometric manuals specifically for architects, such as C. Protot's *Cours spécial d'architecture, ou Leçons particulières de géométrie descriptive* (1838), as well as those for craftsmen, like Charles-François-Antoine Leroy's *Traité de stéréotomie, comprenant les applications de la géométrie descriptive* (1844).[32] Encyclopedic surveys such as *La science des artistes* (1844) compiled contributions by many geometers and architects, summarizing the geometric state of the art. By the last quarter of the nineteenth century, technical manuals evolved into virtuosic showpieces. Louis Monduit's oversized *Traité théorique et pratique de stéréotomie* (1889) presents rigorously discretized solutions to many fundamental geometric problems of nineteenth-century vaults, illustrated in lushly toned drawings.

fig. 2.3 Louis Mazerolle's *Traité théorique et pratique de charpente* (1895) did for carpentry what Monduit's work did for stereotomy, recording intricate projective and stereotomic methods for complex roof structures through lavish illustrations.[33] Meticulous craft, mathematized drawing, and architectural detail fused into a common practice of precise representation.

By the fin de siècle, these geometric techniques had been completely assimilated into design at the technical level of architectural construction. The manuals that encoded them marked an inexorable convergence of geometric technique with building design, but also the advent of a procedurally regimented approach to architectural geometry.[34] The segmentation of mathematical knowledge in geometric manuals ultimately facilitated the encapsulation of that segmented knowledge in a burgeoning armory of complex geometric drawing machines.

The Industrialization of Geometry

fig. 2.4 Drawing machines irrevocably changed the relationship between the architect and the act of drawing. If the architect had been a

craftsman of drawings, the instrument was a factory, and the

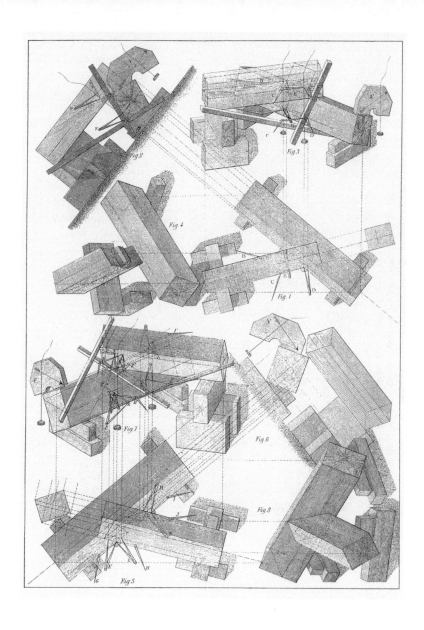

2.3   Stereotomic projections of complex carpentry intersections.
Source: Louis Mazerolle, *Traité théorique et pratique de charpente*
(Paris: Éditions H. Vial, 1895), pl. 16.

2.4 The kinetic traces of circles as they roll within and around other circles, defining epicycloids and hypocycloids. Source: Walther Dyck, *Katalog mathematischer und mathematisch-physikalischer Modelle, Apparate und Instrumente: Unter Mitwirkung zahlreicher Fachgenossen* (Munich: C. Wolf & Sohn), 1892.

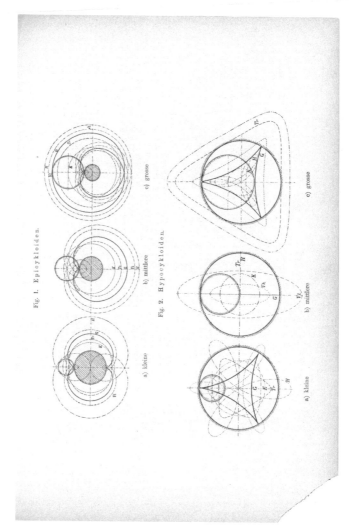

Fig. 1. Epicykloiden.

a) kleine   b) mittlere   c) grosse

Fig. 2. Hypocykloiden.

a) kleine   b) mittlere   c) grosse

architect its manager. The human capacities once so essential to facile draftsmanship were delegated to the drawing machine, supplanted by attention to the logic of the machine itself. The situation was analogous to the change in status of calculation itself in the sciences. Lorraine Daston notes that "by the turn of the nineteenth century, calculation was shifting its field of associations, drifting from the neighborhood of intelligence to that of something very like its opposite and sliding from the company of savants and philosophes into that of unskilled laborers" or, one might add, machines.[35] Yet this reorientation of skills ultimately underwrote new possibilities. Advances in design and instrumentation were often twinned, complementary, and even symbiotic: one enabled the understanding and expansion of the other. Design knowledge and, more particularly, geometric knowledge drove the need for new tools, and these tools in turn sponsored the creation of more advanced design knowledge.

As the virtuous cycle advanced, the demand for ever more precise machinery proceeded rapidly. Many of the earliest prototype drawing machines suffered from manufacturing defects that degraded drawing quality, and advancements in precision followed only after improvements in industrialization. Two manufacturing innovations in particular elevated the quality of drawing instruments. First, the industrial revolution precipitated a host of improvements in accuracy, repeatability, and scale of fabrication that afforded a new level of complexity to machine instruments of all types. A lathe to accurately and reliably thread screws was invented only in 1797, and paved the way for highly accurate and adjustable instruments.[36] The further standardization of machine screws, essential to the development of precise compound machinery, was well under way by the 1830s. Fabrication machines from watchmaking were scaled up to other types of metal manufacturing, dramatically increasing production precision and speed. Greater accuracy and replicability allowed drawing instrument makers to embrace more automated fabrication techniques and expand their market decisively.[37]

Second, the new precision in mechanical components enabled the compound motions necessary for the construction of more complex second-order machines. An example of a first-order machine is a compass. Once the radius is fixed, the compass has one degree of freedom: its angle of rotation. Second-order machines, such as ellipsographs and spirographs, operated with two inter-related compound motions, such as one gear rotating around a second. This second motion increases the required machining accuracy exponentially, since errors in the first motion would be magnified by the second. With more accurately machined components, compound motions became dramatically more precise and reliable. The nineteenth-century advances in fabrication also made possible the wide production of higher-order machines of three or more interlocking motions, based on models like Suardi's geometrical pen. The compounding of motions in a nested and recursive way was a technical and conceptual leap that gave rise to a new generation of instruments.

These machines became an encapsulated knowledge base, each able to rapidly execute the procedures of specific deductive calculations. Through the use of common machines, one discipline could benefit from the solution of related problems in another. Alison Morrison-Low describes the cross-disciplinary consequences of machine advancement, noting that "As problems were solved in one industry, it was immediately realized that the solution was applicable in another where there was a close technical relationship; and it was transmitted to them through the machine tool industry, which 'may be looked upon as constituting a pool or reservoir of skills and technical knowledge which are employed throughout the entire machine-using sectors of the economy.'"[38] Machines became a common, discipline-agnostic interface to underlying geometric knowledge.

A surprising example of this type of methodological diffusion was a series of instruments that translated ideas from differential calculus, such as polar integration, into the mechanical motions of tools to measure the areas of irregular shapes. This class of tools, known collectively as planimeters, was invented by the Swiss mathematician Jakob Amsler-Laffon (1823–1912) in 1854. Based on certain boundary line integrals, Amsler's planimeter could be used to mechanize the area determination of nonstandard figures by simply tracing the outline of a curve. The planimeter would then calculate the area through a series of interrelated counting dials.[39] For architects, the planimeter made nonstandard area calculations, from floor plans to developed surfaces, effortless and accurate. Planimeters enjoyed robust sales among engineers, architects, and planners into the early twentieth century, Amsler himself selling many thousands of units.[40] Amsler went on to develop more sophisticated iterations of this planimeter, including a version that could calculate spherical areas. So compelling were these new mechanical instruments for integration that even the illustrious physicist James Clerk Maxwell (1831–1879), author of the synthetic electromagnetic theory, proposed his own variations on it.[41] Just as differential calculus had sublimated arithmetic operations to continuous analysis, these new machines allowed the control and analysis of a new scope of nonstandard forms and volumes with applications for physics, surveying, and architecture.

These technological developments resulted in a range of machines that each encapsulated specific geometric procedures, each with particular implications for design. Devices such as elliptical and semielliptical trammels automated the design of eccentric arches.[42] The ellipsograph automated the drawing of vault intersections to the point that technical treatises began to recommend the use of the ellipsograph as an essential architectural device.[43] Helicographs encapsulated the surprisingly complex design rules of Ionic columns and certain types of spiral staircases.[44] And conchoidographs captured the equally complex design knowledge needed to exactly describe classically proportioned columns. In a sense, these tools were like custom software applications for each of these specific design problems. They augmented the forms conceivable by the architect and systematized the use of complex curves within design.

fig. 2.5

At their core, drawing machines were calculating devices with a graphic output. But a lack of numeric output did not diminish their relevance to industry. Quite the contrary: the sophistication of graphic calculation was ideally suited to some of the most complex computational problems of the Victorian era. The porosity between drafting, calculation, and scientific instrumentation was exemplified by Joseph Clement (1779–1844), a well-known draftsman and one of the foremost makers of precision instruments in nineteenth-century Britain.[45] Clement won national recognition for the design of mechanical drawing instruments, particularly those for the construction of ellipses, and he

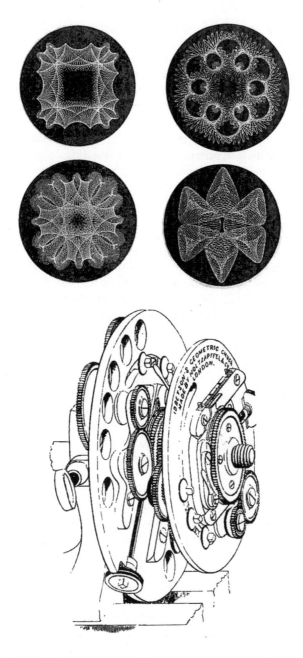

2.5   A geometrical lathe or rose engine, with instances of curves produced by its permutations.
Source: Thomas Sebastian Bazley, *Index to the Geometrical Chuck* (London: Waterlow and Sons),
1875, frontispiece and plates.

worked extensively on the manufacture of automatic machine tools.[46] In 1824, when Charles Babbage (1791–1871) needed a mechanician to execute his massive difference engine, he enlisted Clement's expertise.[47] Babbage's difference engine, which was designed to calculate extensive logarithm tables, was the most ambitious and systematic attempt at automatic computation up to that time. As the principal engineer behind its execution, Clement drafted and coordinated the interworking of hundreds of high-precision components. His expertise in complex curves and the engineering of drawing machines was an indispensable basis for the realization of the difference engine. Ironically, this watershed of automatic arithmetic relied on the geometric calculations of drawing machines. Yet the difference engine, with other devices like Amsler's planimeters, was part of a larger trend to mechanize calculus-based computation through the geometric intermediary of the drawing. Drawing and calculation advanced at lockstep through common mechanical instrumentation.

The knowledge underlying geometric constructions as well as the mechanical means for their execution forged a new symbiosis between design and geometric technique in the nineteenth century. Through the expansion of geometric understanding of second-degree curves, such as conic sections, and of surfaces, such as ruled and extrusion surfaces, architects, engineers, scientists, and draftsmen created a new demand for what were essentially computation machines for the facile drafting of geometric forms. Increasingly sophisticated machines in turn were put to use in the drawing machine industry itself, creating another virtuous cycle of innovation resulting in yet more sophisticated machines. In a *mise en abyme* of instrumentalization, the knowledge captured by the machine could only be accurately encoded by using another machine (a phenomenon with which we are even more familiar today). The reciprocity between machine innovation and formal innovation precipitated a remarkable intricacy in the architecture of the nineteenth century's twilight.

Machine as Media

Architecture reveals itself in the dance between culture and technique, and drawing instruments—mechanical and digital—

serve as intermediaries through which disciplinary intuitions are concretized in discrete drawings and forms. Design and instrumental knowledge are symbiotically conjoined. As the anthropologist David Mills observed, the content of a discipline is "social as much as intellectual, psychic as much as political, ethical as much as methodological."[48] Yet the boundary between design and instrumental knowledge is porous: what begins as technical vocabulary may migrate to cultural terms of art, while cultural values inevitably influence the adoption of technical systems. Machines and instruments are media through which architects access both mathematical techniques and cultural priorities. For instance, as new geometric methods appeared in the nineteenth century, they were ultimately encapsulated in particular drawing and measuring machines. Embedded within these machines, geometric knowledge could be simulated by the user through instrumental machine use. Advanced mathematics was thus smuggled into design culture through the back door of instruments that black-boxed underlying methods.

Once design knowledge is encapsulated in a machine process, access to it becomes dependent on both instrumental knowledge of the machine and the machine itself. The tool is essential to the act of design. If the instrumental knowledge of tool operation is lost, access to the associated processes themselves is also lost. Instrumental knowledge becomes integral to architecture itself.

Instrumental knowledge is a vital but fragile disciplinary capacity that can be lost with cultural vicissitudes of designers' attitudes, tastes, and fashions. Intellectual shifts in early twentieth-century European design jettisoned not only the mannered ornamentation of the late nineteenth century, but also the rich technical culture that attended it. Forms that were simple to reproduce were preferred to technical virtuosity.[49] In the embrace of economical forms, the expertise to create complex shapes – including the practices of mathematized drafting and the associated drawing instruments – receded from design practice.

This epistemic, formal, and technical breach had lasting consequences. Ambitious geometric experimentation was virtually absent from architecture for nearly half a century. Ruled surfaces, already subjects of explicit study by Douliot as early as 1825, did not reemerge in the architectural vocabulary until the 1950s, with the hyperbolic work of Félix Candela or Miguel Fisac, and then without the aid of advanced geometric drawing machines. Even sophisticated protocols such as the cartographic subdivision of Buckminster Fuller's geodesic domes were undertaken, at least initially, without the advantage of complex drawing instruments. These architects found themselves reinventing the craft of curved form in architecture, rediscovering geometric design knowledge that would have been enabled by a forgotten technology. The loss of instrumental knowledge had foreclosed the highest levels of geometric control and sophistication.

This rupture of continuity and loss of knowledge was not inevitable. If instrumental knowledge and its formal implications had been embraced as an integral aspect of the culture of architecture, the formal limits of twentieth-century modernism would have been mitigated by an unbroken trajectory of technical innovation borne in technical instruments. In other words, if instrumental knowledge had been seen as a fundamental part of architectural knowledge, the machine culture of architecture could have been properly understood as a part of the continuous evolution of the practice of architecture itself. Our current computational culture would have an integral history and epistemology stretching back to the nineteenth century.

Encapsulated mathematical processes are again transforming architecture, today through the digital instruments of computer software. The implications echo those of nineteenth century drawing machines: design by instrument opens rigorous geometric possibilities through a mastery of technical methods, which in turn require a reorganization of design pedagogy and, ultimately, architectural knowledge. Tools again become a common medium between design, engineering, and science. Yet, like drawing machines, digital software dramatically increases our reliance on representational and operational systems that we incompletely understand but nevertheless trust implicitly. Antoine Picon has noted that "mathematical principles are very often hidden behind their effects on the screen. In many cases the computer veils the presence of mathematics."[50] Yet even hidden in black-boxed software, encoded mathematics furnishes a foundation for geometric innovation.

57

Ways of encoding the technical knowledge of design, including in drawing machines, geometric manuals, and digital software, are essential to the culture and practice of architecture. Without them, the highest registers of controlled spatial and geometric virtuosity are lost. Yet there has been a strong tendency, arguably since Alberti, to dichotomize design knowledge and instrumental knowledge, and to relegate technical or mechanical expertise to the domain of specialists or operators. Perhaps this is due to a mistrust of black-boxed mechanical, electrical, computational, or conceptual operations of which the architect cannot have complete understanding, or simply an assumption that the culture of architecture resides elsewhere.

Knowledge that is instrumentally encoded actually represents a radical new kind of collective cultural memory, a vast cabinet of new disciplinary practices. With it, design practices can be made modular and accessible through tools: they can be abstracted, encoded, trusted, mechanically or electronically represented, and propagated. A new epistemology of automatic and recombinant knowledge is possible through instrumental encapsulation. Alfred North Whitehead aptly argued for the cultural power of knowledge encapsulation: "It is a profoundly erroneous truism, repeated by all copy-books and by eminent people when they are making speeches, that we should cultivate the habit of thinking what we are doing. The precise opposite is the case. Civilization advances by extending the number of important operations which we can perform without thinking about them."[51] The amalgamated artifacts of embedded knowledge, from drawing instruments to digital procedures, comprise vast *Wunderkammern* that align mathematics and architecture in a rich machine epistemology.

3 Theorems Made Flesh: The Architectonics of Mathematical Maquettes

3.1   A typical excerpt from a commercial catalog of mathematical models displaying a range
of wares.   Source: *Verzeichnis mathematischer Modelle* (Berlin: Sammlungen H. Wiener
und P. Treutlein, 1912), table v.

The exact sciences of the nineteenth century were animated by a fig. 3.1 feverish search for tangibility. Perhaps in reaction to the increasingly intangible frontiers of Belle Époque physics, chemistry, and mathematics—magnetic and electrical fields, microscopic molecular structures, non-Euclidean geometries—the creation of tangible physical maquettes rendered spatial intuition concrete in scientific practice. Invisible entities from abstract geometric surfaces to chemical arrangements began to appear as wood and plaster models in the scientific spaces of nineteenth-century Europe and North America. Indeed, a cottage industry of craft within the exact sciences grew up around the peculiar and specific requirements of making scientific models. Through physical representations of theoretical entities, model-making practice recast how the forms of science were intellectually and aesthetically consumed. Beyond their explanatory value within scientific practices, models molded a public image of scientific concepts and how broader cultures might imagine them.

In arguing for the scientific maquette as a type of knowledge, the philosopher Davis Baird claimed "models do much of the same epistemological work as theories," including functioning as tools of common understanding.[1] They are objects to reason about, with, and through. Most critically, they are public artifacts of collective reflection and persuasion. They are created for classrooms, exhibitions, and widely distributed catalogs. Beyond their illustrative role, they are social objects of debate, consensus, and imagination.

While they often acted as visible pedagogical instruments for chemistry or biology, physical models gave corporeal presence to mathematics, arguably the least tangible of the exact sciences. They allowed theorems to slip into skins and become not only products of calculation, but cultural organisms in their own right. First in early nineteenth-century France and later in imperial Germany, vibrant model-making and model-consuming subcultures evolved to communicate the strange and astonishing geometries of differential and non-Euclidean spaces. For mathematicians and later for architects and artists, these models served the double role of both wondrous exemplars of method and objects of aesthetic possibility. Models articulated how advanced mathematics should look, as practice and idea. As totems which exuded a mystery of cryptic knowledge, physical scientific models endowed their uses with commensurate mystique.

As science searched for tangibility, architecture hunted for transcendence. In the work of Enlightenment designers like the French architect Jean-Nicolas-Louis Durand (1760–1834) there was a restless quest for abstract systems that could elevate building ideas beyond the idiosyncrasies of individual projects. It was in the common practice of model making where the incarnation of mathematics and the abstraction of architecture crossed paths. Mathematical maquettes could be possessed—mentally and physically—and thus were co-opted as material proxies to circulate the form and content of mathematical ideas into architecture in a ready-made way. From the classroom to the studio, geometric maquettes shaped a modern architectural intuition of form. In some very literal cases, mathematical models were simply scaled to the size of a building to become architecture. In other cases, the logic underlying the models became a template for architectural creation. Regardless of how they were deployed, two major species of nineteenth-century mathematical models—string-based ruled-surface models and models of differential and topological surfaces—served as catalyzing ur-forms for distinct formal and calculational cultures within architecture. Linear ruled-surface models were hyperbolic architecture in microcosm, while plaster models of differential warped surfaces invited intricate topological experimentation. Each of these two key model types was diffused into architectural culture along slightly different vectors, and each will be revisited in later chapters. Here we only hope to trace the initial paths of consumption, propagation, and impact of scientific model culture on design. Both as methodological exemplars and formal archetypes, models lodged a panoply of marvelous new elements in the aesthetic lexicon of architecture.

String Instruments

Mathematicians have long employed demonstrative physical maquettes to elucidate and inspire. Simple wooden forms of conic sections or Platonic solids are time-honored props for the explanation of basic geometry. As geometry evolved in new and exotic directions in the nineteenth century, model making developed as a corollary of those innovations, physically concretizing forms which otherwise might remain algebraic or differential abstractions. Maquettes were geometric messages in a bottle, signals transmitted from the hinterland of mathematical imagination.

Modern models were injected into the heart of nineteenth-century engineering and architectural training by the polymathic French engineer and educator Gaspard Monge (1746–1818). With France in the throes of national competition with rapidly industrializing England, Monge's ambition was to train a new type of engineer with a spatial intuition equal to the challenges of designing machinery to surpass the quality of the then-dominant British workshops.[2] A new Paris engineering school, which he participated in founding in 1794 and which became the famed École Polytechnique, was the St. Peter's of this new religion, and Monge was its pope. Engineers needed to be able to mentally orchestrate elaborate assemblies of precise parts, and the keystone of this new mental facility was a training of vision. To this end, Monge developed a series of drawing practices he termed *géométrie descriptive*, the forerunner of modern orthographic drawing, a system he promoted as a "universal language of industry."[3] Using paired planes of projection and intricate rules for constructing the intersections of lines in space, descriptive geometry provided a system for exactly calculating complex spatial relationships, from engineered machinery to architectural geometry. Monge was equally conversant in both complex mathematical developments and the spatial problems of design. These paired interests intersected in his study of *surfaces réglées*—ruled surfaces. Monge's work paralleled the earlier work of the mathematician Leonhard Euler (1707–1783) on surface developability—that is, the unrolling of spatial surfaces onto flat planes without distortion. Euler had pioneered the study of developable surfaces in his 1772 paper "De saidis quorum superficiem in planum explicare vicet" (On solids whose surface may be spread out upon a plane).[4] Monge independently articulated notions of ruled and developable surfaces and applied these ideas to problems of mechanical engineering and architecture, such as gear motions and vault intersections.[5] The ruled surface, in which each point of the surface is on a line that also itself lies entirely on the surface, was a particularly critical construction. The characteristic curves of the ruled surface, the so-called ruling lines, drew a thin net of fibers tautly stitched in tensioned spatial membranes.

Monge was a spirited advocate of physical visualization, and ruled-surface models were among the most beguiling physical products of his approach. He commissioned the creation of fine thread models by his student Théodore Olivier (1793–1853) to illustrate principles of ruled surfaces, developability, and descriptive geometry.[6] Olivier's models were delicately appointed, with hundreds of thin wires attached to brass frames fixed to hefty wood bases. Each thin wire represented a ruling line. Arrayed in precise series, the threads were weighted with individual plumbs, and multiple wires were bound together at their intersection points by delicate brass rings. Some models were hinged and movable, kinetically demonstrating the dynamic intersections between ruled surfaces shifting in space. Because the mathematical models presented geometry as both an analytical and a spatial concept, they signaled an emerging intuition in which calculation could be registered in physical and spatial dimensions.

Monge and Olivier's hinged models were akin to devices for mathematical experimentation, existing somewhere between a static maquette and a mechanically dynamic drawing instrument. Kinetic models physically calculated whole families of surfaces quickly, reliably, and accurately. They were a way to see a process of calculation at work, to intervene in it at will, and to associate it directly with a range of forms. Wire models were not static archetypes but instead mechanical systems for inductive extrapolation. They were a physical manifestation of mathematical proof.

Olivier crafted his meticulous ruled models prolifically, and engineering schools were ready consumers.[7] At first idiosyncratic artifacts of the École Polytechnique, string models rapidly became standard collateral for engineering courses across European and the United States in the latter half of the nineteenth century. Olivier initially fulfilled much of the early demand, but wireframe models soon blossomed into a global phenomenon. They found homes in engineering and scientific institutions such as the Conservatoire Nationale des Arts et Métiers, Harvard, West Point, Cornell, Union College, and many other research institutes and universities. The globalized model market signaled a newly vital visual culture of mathematics.

fig. 3.2

fig. 3.3

3.2 An example of an Olivier-type ruled-surface string model.
Source: Collection of Historical Scientific Instruments, Harvard University.

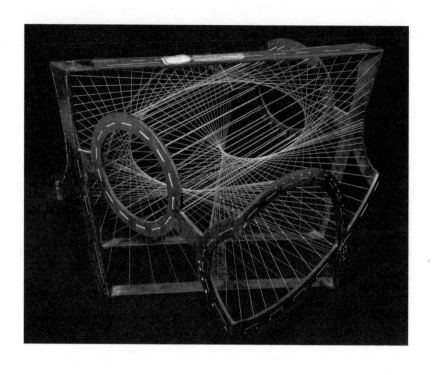

3.3 A string model with rigid frame in the archives of the Institut Henri Poincaré,
late nineteenth century.   Source: Photo courtesy the author.

Calculated Bodies

Ruled-surface models were only the first foray into a vast wilderness of sculpted mathematical forms, and mathematicians avidly consumed new models of ever more exotic surfaces. Toward the fin de siècle, sophisticated plaster models of complex algebraic, projective, and differential surfaces documented the furthest frontiers of science. In mathematical pedagogy, the entire sensibility around mathematical models also evolved. Where wire models bounded void volumes in outline, plaster models had the firm bodily presence of classical statuary. Traced with regulating curves and cast in stark bone-white, their elegant presence attracted admirers among photographers, sculptors, visual artists, and designers. Biological associations were inescapable, due to their smooth skin, shell-like silhouettes, and muscular inflections. More than representations, they were embodiments. They were formulas in the flesh, poised to twitch, leap, or pirouette in place. Moreover, plaster surface models were not limited to the ruled geometries of wire models. They could incorporate robustly curved features belonging to other, nonruled surfaces. Beginning around 1870, the deepest and most modern research on the properties of polynomial surfaces of degree higher than two could be rendered incarnate through plaster models.[8] In such surfaces, with dramatic singularities and violent cusps, the malleability of space itself was on turbulent display.

Whereas Olivier's string models were instruments for manipulation, plaster models were more akin to the biological samples of the naturalist. These fossilized bodies invited anatomical analogies. Yet they were as much maps as objects, often crisscrossed by lines of demarcation, isoparametric curves, and other graphic meridians. Due to their sheer volume, variety, and quality, they came to exemplify the whole genre of mathematical maquette in the late nineteenth and early twentieth centuries.

The German mathematician Felix Klein (1849–1925) elevated the use of scientific maquettes to new heights of both rigorous research and public visibility. Perhaps best known to architects as the originator of the famed Klein bottle, Klein was a towering figure of fin-de-siècle mathematics, a great unifier who applied algebraic techniques to reveal profound geometric structures.[9]

fig. 3.4

3.4   Edvard Neovius's triply periodic minimal-surface model, among the first to explore and
physically model infinitely extensible minimal-surface geometry.
Source: Edvard Rudolf Neovius, *Bestimmung zweier speciellen periodischen Minimalflächen, auf
welchen unendlich viele gerade linien und unendlich viele ebene geodätische Linien liegen*
(Helsinki: Frenckell, 1883), pl. IV.

In reaction to the purists who then dominated German mathematics, Klein envisioned an expansive and cross-disciplinary role for mathematics that encompassed the applied sciences. In this context, Klein saw the unique virtues of maquettes that could serve as common points of reference between mathematics and engineering. From a theoretical point of view, he understood specific models as instances of larger families of surfaces, elements in a kind of comparative anatomy of calculation. Yet, for Klein, models were not merely esoteric curiosities. As a relentless educational reformer, he saw the imagination for envisioning complex spatial operations that models could sponsor as a crucial pedagogical aim.[10] The vivid presence of a physical model was an incomparable aid to space perception for students at every level.[11] Klein inherited his visual disposition from his PhD advisor Julius Plücker (1801–1868), a renowned mathematician in his own right, who first introduced Klein to the advantages of physical models in the 1860s.[12] Klein was also influenced by an exhibition at the Arts et Métiers in Paris that included Olivier's *surface réglée* string models in 1870.[13] In the ensuing years Klein would transform not only the disciplinary dimension of visual mathematics but how the broader culture consumed mathematical objects.

When Klein arrived at the Technische Hochschule in Munich in 1875, he established the nucleus of his bold new ecology of visual mathematics. From an architectural perspective, a fortuitous collaboration fueled Klein's project: the presence of erstwhile architect Alexander Brill (1842–1935). Brill shared Klein's commitment to reforming teaching and research through visual maquettes, and moreover Brill had the rare hybrid background to make that possible. He originally trained as an architect at Technische Hochschule in Karlsruhe and University of Giessen, graduating in 1864. Attracted to innovations in non-Euclidean and hyperdimensional geometry, he completed his doctorate in mathematics with a dissertation on curves with hyperelliptic coordinates. In 1874, he began crafting early cardboard surface models, applying both his mathematical expertise in model making and his architectural skills in analytic drawing. At the Munich Technische Hochschule, Klein and Brill enthusiastically introduced plaster models to aid the explanation of higher-dimensional geometry.[14] The techniques used to fabricate these models were characteristic of plaster architectural or sculptural models of the period—a Corinthian capital or fragment of statuary drapery would have been executed the same way. Most commonly, frameworks of crisscrossing cardboard slices or wire guides were assembled and then deftly finished in plaster.[15] Another common construction was a layered topography of contoured wood slices, plotted from functional values or implicit thresholds, finished with plaster of Paris.[16] On occasion, delicate models were directly cast in plaster by contracted expert craftsmen. In the bodies of maquettes, mathematics was manifest not only as a corpus of knowledge but also as a corporeal product.

fig. 3.5 Through the gradual accumulation of a prodigious archive of mathematical models, Klein and Brill shifted the role of the individual model from a singular curiosity to an integral member of a larger taxonomized collection. Each model was thus subsumed in the theoretical framework of surface classes that the archive or collection reflected. Such a collection also implicitly organized a social project of model production. As Klein's modeling project escalated, he established a model construction laboratory and required dissertations on algebraic surfaces to produce models that were duly deposited in his prodigious archives.[17] The crafting of a new model, paired with the discovery of each new kind of surface, became a rite of initiation into the elite world of research mathematicians. The collection became a plan for and evidence of the creative and social project of mapping the universe of mathematical form.

With Klein's blessing, Brill and his brother Ludwig set up a factory and marketed the models as pedagogical aids through their family's publishing company. Alexander Brill provided the templates for the models, which were photographed for widely distributed catalogs. In 1899, Martin Schilling purchased the factory from the Brills and further expanded the scope and scale of their distribution, moving it from Darmstadt to Halle and finally Leipzig.[18] From the mid-1870s until 1932, the Brill-Schilling factory provided the raw material to massively expand the popular acquaintance with forms at the frontiers of complex geometry.[19]

By the 1890s Brill's spark had ignited an explosion of imitators, competing at smaller scales in Europe and the United States. In America, Richard P. Baker (1866–1937)[20] and Albert Wheeler

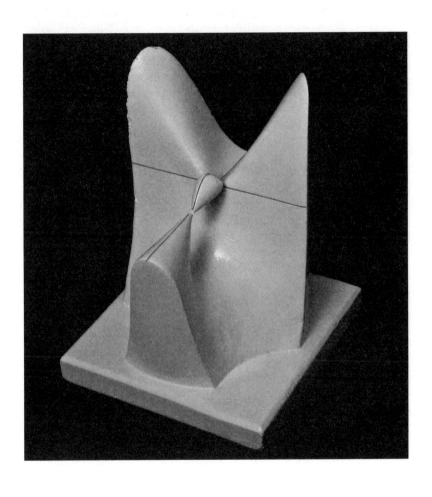

3.5 A partially ruled implicit surface, typical of the Brill-type plaster maquettes.
Source: Photo courtesy the author.

(1873–1950) produced their own models for distribution, in alignment with a new emphasis on visual methods of teaching promoted by the Mathematical Association of America.[21] After Klein's example, some universities, such as the University of Illinois, even built their own workshops for creating mathematical models. By the early twentieth century, these models proved so phenomenally popular that they had their own generously illustrated entry in the *Encyclopedia Britannica*, complete with detailed categorizations of their sundry varieties.[22] A collection of mathematical maquettes became the sine qua non of progressive scientific and engineering education, a signal within the academy of a view toward the future.

Computational Geometry of the Belle Époque

fig. 3.6 Architecture and engineering education were imbricated in turn-of-the-century Europe and America, and many schools offered joint mathematics courses for engineers and architects in which geometric models played a vital role. The models used for mathematics, engineering, and architecture courses were rich mélange of technical representations. For instance, some of the models and drawings in the archive of Klein's collection at the University of Göttingen are in fact explicitly architectural, like the masterful projective derivation of intricately folded mansard roofs. They attest to the porosity between disciplines at that moment. Mathematical models of newly discovered differential surfaces sat side by side with models, images, and stereograms of more practical descriptive geometry, cohabitating in the same ecumenical space of representation. Until the close of World War I, architects could encounter wire, plaster, and perhaps even kinematic models as a matter of course in their academic training.

Since architecture and engineering education were conjoined, Monge and Olivier's models circulated internationally in late nineteenth-century architectural circles. Monge's pedagogical influence on architecture is legible, for instance, in the education of the famed Catalan architect Antoni Gaudí (1852–1926). As Mark Burry, Jordi Coll Grifoll, and Josep Gómez Serrano have noted, Gaudí was first exposed to Monge's ruled surfaces through a course on descriptive geometry taught by José Castelaro y Saco.[23]

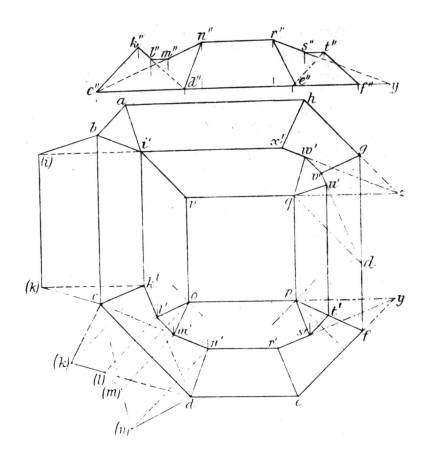

3.6 Plan and elevation of a descriptive development of a complex roof structure.
Source: Collection of Mathematical Models, University of Göttingen.
© Collection of Mathematical Models, University of Göttingen.

Ruled surfaces became a fundamental vocabulary of Gaudí's constructions, particularly his Temple Expiatori de la Sagrada Família in Barcelona. In that prodigious confection, the deployment of ruled surfaces borders on riotous. Grand gestures as well as subtle details revel in a virtuosic play of ruled constructions. Among many, one particularly striking example is in the frames of certain windows, ruled hyperbolas of revolution, whose intersections were calculated as projected ellipses.[24] The drawing and modeling of these apertures would have rested on techniques from Monge and Olivier, and even Gaudí's monumental hanging-chain structural models owe considerable debt to Olivier's string maquettes. For Gaudí, as for many others, Monge's approach to complex surfaces was an inherently architectural innovation.

fig. 3.7

Yet Gaudí's facile use of ruled surfaces was far from the only architectural manifestation of these techniques in the late nineteenth century. In fact, ruled-surface models were used to describe many types of architectural details, as is clear from a description of the Fabre de Lagrange ruled-surface models of the Science Museum, London:

> By far the most interesting models ... are those which show the connection of the cone and cylinder with their tangent planes. Scarcely less interesting are the models of the conoids and the ruled surfaces with director planes ... a very remarkable example of which is the French skew arch called the *biais passé* ... the object of which is to connect two openings not exactly opposite each other and not necessarily parallel to one another.[25]

These descriptions were no mere speculations. A vaulted passage in the Apostolic Palace at the Vatican, in which a ruled surface connects two entrances in planes skew to each other, illustrates explicitly the *biais passé*. The strange situation of the two adjacent rooms, slightly displaced from each other, induces this precisely torqued form. Mathematical models thus demonstrated concrete solutions to the eccentric puzzles of nineteenth-century space-making.

Architects opportunistically drew on this menagerie of mathematical curios to imagine the training of the designer in new ways. Mathematical models were employed not only as

3.7   A *biais passé* ruled-surface skewed arch in the Vatican.
Source: Photo courtesy the author.

illustrations of tectonic strategies but as explanations of projective technique and as tests of drawing. A course description from Princeton's 1889 catalog elaborates on these uses:

> The drawing courses in descriptive geometry, ... [and] architectural constructions ... are illustrated by a large collection of models, which includes a number of duplicates of the Olivier ruled surface models, two hundred of the Schroder (Darmstadt) manufacture, complete sets of the mathematical models designed by Professors Brill and Bjoring, a number of the "Muret" plaster models, and several warped-surface models from designs by the professor.[26]

Skiagraphic practices—that is, exercises for rendering projective shadows—were another common use of these models. Mathematical models were the ideal objects of such drawings. The geographic range of their diffusion is evident in the fact that Frank Lloyd Wright (1867–1959), for instance, meticulously drew, using descriptive geometry, the shadows of surfaces of revolution during his own formal training in design at the University of Wisconsin–Madison. Like innumerable other architects-in-training of the age, Wright encountered geometric maquettes as an organic aspect of the hybrid architect-engineer's visual education.

Atlases of Warped Form

Over time, models migrated from the classroom to public exhibition halls, fairs, and museums. As they did so, they began to take on broader cultural associations that stretched beyond their mathematical origins. In this translation they moved from objects of research to objects of communication, and became popular signs and commodities of advanced mathematics and engineering. After 1920, architects in the United States and Europe encountered mathematical maquettes less often in disciplinary contexts and more frequently in events popularizing scientific ideas or in cataloged collations of model images that designers adapted as sourcebooks. In fact, the generation of designers of the early twentieth century encountered models only in these more eclectic contexts, which freed architects to interpret them in less prescriptive and more culturally inventive terms. Architectural encounters with these uncanny maquettes included public exhibitions and at least two types of texts: illustrated manuals of engineering geometry and commercial catalogs of mathematical models.

The physical collections of models mounted in public and private exhibitions were an alternate mode of indexing modern mathematics. While the vitrines accumulated by mathematics departments familiarized the scientific elite with these forms, the public at large became acquainted with mathematical models in part through world exhibitions. For example, the directory of Chicago's expansive 1893 Columbian Exhibition included listings for at least five mathematical instrument and model makers from Sweden, Germany, and the United States.[27] Models thus became signs of national distinction. Objects of wonder and fascination, they took their place alongside the other prodigies of labor at the fair, packaged for consumption by an awed public.

The exhibition altered the status of models yet again. They became emblems of national culture and indices of science wrestling with problems at the edge of visual intuition. Of course, exhibitions were also places of commerce, and models were there to be sold. In moving from the mathematical archive to the exhibition, models also transformed into ready-made industrial and cultural commodities.

In contrast to public exhibitions, manuals and catalogs of mathematical models were intimate portable codices of geometric research. Often rich with elaborate drawings, manuals and catalogs served as de facto surveys of the scientific state of the art and sourcebooks for surreal new visual idioms. Consider, for example, the work of German mathematicians Eugen Jahnke (1861–1921) and Fritz Emde, whose 1909 *Funktionentafeln mit Formeln und Kurven* was a mammoth index of warped surfaces and the functional algebras that undergirded them. Brimming with engaging illustrations of mesmerizing new objects, *Funktionentafeln* anatomized mathematics, dividing the species of models with a lepidopterist's eye. It served as a valuable visual reference for engineers and mathematicians of the early twentieth century.

Jahnke and Emde's manual also had surprising currency with artists and designers, for whom such manuals became files of uncanny objets trouvés. The surrealists already had a marked affinity for mathematical objects as readymades. In 1936, in a famous exhibition at Charles Ratton's Gallery, the French writer André Breton (1896–1966) organized an eclectic mélange of items ranging from natural objects of the mineral, vegetable, and animal kingdoms, various ready-made industrial products, Native American and Oceanic artifacts, and of course mathematical maquettes.[28] In this context, the models were placed in the provocative and free associative milieu of the *cabinet de curiosités* as one among many comparable aesthetic objects.

figs. 3.8–10

A paradigmatic case of the use of mathematical forms as surreal found objects is German artist Max Ernst's (1891–1976) exquisite 1948 collage *La fable de la souris de Milo*, an image of the Louvre's Venus de Milo grafted to a differential surface taken directly from Jahnke and Emde's catalog of geometric surfaces. The selected surface was a Hankel function, an integral function across the complex plane that included both real and imaginary numbers, evoking also the imaginative and transgressive nature of the collage itself. The conjunction of this surface with an image of classical statuary made for a jarring but suggestive nexus of speculative and tangible realms.

Ernst channeled an omnivorously eclectic sensibility, and his opportunistic sampling from various types of scientific catalogs was an essential part of his process. For him, the catalogs were gateways to new modes of sight and knowledge. He recounted,

> One rainy day in 1919 in a town on the Rhine, my excited gaze is provoked by the pages of a printed catalogue. The advertisements illustrate objects relating to anthropological, microscopical, psychological, mineralogical and paleontological research. Here I discover the elements of a figuration so remote that its very absurdity provokes in me a sudden intensification of my faculties of sight—a hallucinatory succession of contradictory images, double, triple, multiple, superimposed upon each other with the persistence and rapidity characteristic of amorous memories and visions of somnolescence. These images, in turn, provoke new planes of understanding.[29]

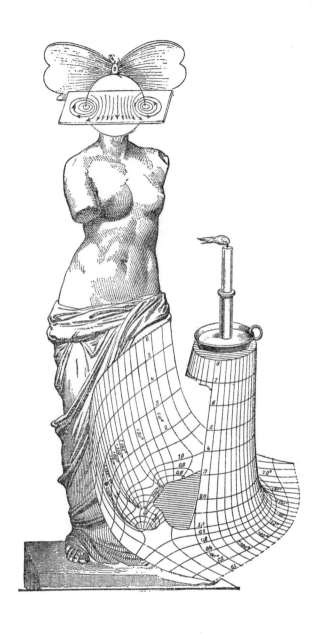

3.8 Max Ernst, *La fable de la souris de Milo*, 1948.
Source: The Copley Galleries, *At Eye Level: Paramyths*, exhibition catalog,
© 2020 Artists Rights Society (ARS), New York / ADAGP, Paris.

d) Funktionen dritter Art $H_p^{(1)}(x)$ und $H_p^{(2)}(x)$.
d) Functions of the third kind $H_p^{(1)}(x)$ and $H_p^{(2)}(x)$.

Für reelles Argument haben diese von Hankel eingeführten Funktionen komplexe Werte. Dagegen sind die Ausdrücke

These functions introduced by Hankel have complex values for a real argument. But the expressions

$$i^{p+1} H_p^{(1)}(iy) \quad \text{und} \quad i^{-(p+1)} H_p^{(2)}(-iy)$$
$$\text{and}$$

reell für positives y. Die Bedeutung der H-Funktionen für die Anwendungen liegt vor allem darin, daß unter den Zylinderfunktionen sie allein für unendliches komplexes Argument verschwinden, und zwar $H^{(1)}$, wenn der imaginäre Teil des Arguments positiv, $H^{(2)}$, wenn er negativ ist:

are real when y is positive. The H-functions owe their importance for applications to the fact, that among the Bessel functions they alone vanish for an infinite complex argument, viz. $H^{(1)}$ if the imaginary part of the argument is positive, $H^{(2)}$ if it is negative:

$$\lim_{r=\infty} H_p^{(1)}(re^{i\vartheta}) = 0, \quad \text{wenn}$$
$$\lim_{r=\infty} H_p^{(2)}(re^{-i\vartheta}) = 0, \quad \text{if} \quad 0 \leqq \vartheta \leqq \pi \quad \text{ist.}$$

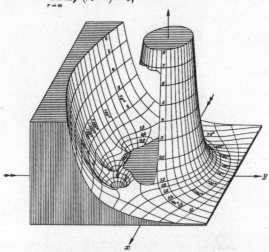

Fig. 73. Relief der Hankelschen Funktion $H_{3,5}^{(1)}(z)$.

Fig. 73. Relief of the Hankel function $H_{3,5}^{(1)}(z)$.

3.9   Relief drawing of the Hankel function.
Source: Eugen Jahnke and Fritz Emde, *Tables of Functions with Formulae and Curves*
(New York: Dover, 1945), 133.

189. Aphrodite (Venus) från Melos.

3.10   Aphrodite (Venus) de Milo, from Ernst Wallis,
Illustrerad Verldshistoria (Chicago: Svenska Amerikanaren, 1894), 518.

Scientific catalogs satiated only a small part of Ernst's voracious appetite for new visions. His other sources included illustrated histories of various kinds, including the world history of the Swedish historian Ernst Wallis (1842–1919), from which the Venus de Milo of *La fable de la souris de Milo* was apparently cut. The embodied dialectic of those two fragments in conversation exposes the fresh conceptual sympathies and conflicts of adapting mathematical forms for broader cultural ends.

La fable de la souris de Milo bears all the tensions and dualisms of the relationship between art and science in a single strange image. The uncanny confrontation of a classical mythological statue and a modern mathematical surface makes those polarities tangible and specific.[30] From their roughly identical proportions, one may be tempted to infer an equality between the statue and the surface, or between art and mathematics. Apparent parallels in representation appear to reinforce an affinity between the two. Reportorial linework—a nearly mechanical hatching for the statue and a network of curves for the mathematical surface—betrays a certain documentary detachment toward both elements. Both are curved and inflected, suggesting a formal sympathy of lively and dynamic shape. Both statue and surface are cast as archetypes, primal forces fused in concert.

Yet, despite the implied equilibrium between statue and surface, *La fable* rumbles with paradoxical tensions. Is the statue-surface an object of both art and science, or neither? Is it more monument or method? Are humanism and technique reconciled in it, or locked in perennial contest? Is the antiquity of the statue primary, or rather the primordial antiquity of mathematics itself? Is the impermanent human presence melting into ageless mathematical abstraction, or is the human emerging from the mathematical? Whatever equilibrium exists between the two seems unstable, a singularity on the verge of tumultuous transformation. With *La fable*, Ernst offers an early twentieth-century vision of mathematical figuration grappling with all of the potencies of contradictions that would come to define the dialectic between mathematics and cultural creation.

Classificational Catalogs

The catalogs that Ernst so enthusiastically drew from were microcosms of the obsessive search for formal categories that consumed mathematical culture of the day. The elusive holy grail was a theorem to schematize and map the familial relationships among the apparently inexhaustible varieties of bizarre surfaces. As David Rowe has observed, "These geometric investigations were pursued almost like botanical studies in which the geometer went about collecting various specimens associated with certain classes of equations. These then had to be classified according to some larger scheme, certainly less ambitious than the Linnaean framework, but nevertheless with a similar purpose in mind."[31] Klein himself made significant strides through his promulgation of the so-called Erlangen Program of 1872 to classify surfaces through advanced algebras. The indexed catalogs of mathematical models were imprinted with the same thirst for taxonomic order.

Mathematical texts intended for a general audience echoed the taxonomic proclivities of model catalogs. Sitting somewhere on the spectrum between technical manual and popularization was the 1932 book *Anschauliche Geometrie* (translated into English in 1952 as *Geometry and the Imagination*) by the mathematicians David Hilbert (1862–1943) and Stephan Cohn-Vossen (1902–1936). Hilbert, one of the preeminent mathematicians of the turn of the twentieth century, spared no rigor in his ambition to "give a presentation of geometry, as it stands today, in its visual, intuitive aspects. With the aid of visual imagination, we can illuminate the manifold facts and problems of geometry."[32] Indeed, *Geometry and the Imagination* was generously illustrated with careful line diagrams and arresting images of mathematical models that Klein had so enthusiastically advocated at Göttingen. That should come as no surprise, since Hilbert joined Klein at Göttingen in 1895 and embraced an approach that was resolutely model-centric: he advocated "instead of formulas, figures that may be looked at and that may easily be supplemented by models which the reader can construct."[33] These meticulous diagrams are presented in the service of an equally rigorous classification of mathematical entities ranging across crystallographic lattices, kinematic contraptions, and differential surfaces. The book also introduces its

fig. 3.11

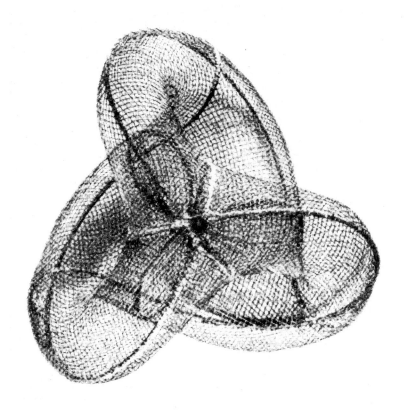

3.11   Photograph of a mathematical model of a projective surface
from Hilbert's *Anschauliche Geometrie*.   Source: David Hilbert and Stephan Cohn-Vossen,
Geometry and the Imagination (New York: American Mathematical Society, 1952).
© 1996, Springer-Verlag Berlin, Heidelberg.

lay audience to advanced methods of topological surface construction, including the joining of edges of elastic sheets successively to develop more elaborate enclosures, illustrated with cloudlike wire-mesh models.

Geometry and the Imagination was a cultural phenomenon which planted a specific type of geometric intuition within the popular imagination at large and in architecture in particular. It was widely cited by architects with a mathematical inclination such as Christopher Alexander (b. 1936) and Lionel March (1934–2018). Texts like it short-circuited scientific training by offering the mathematical models directly as design objects, unencumbered by theoretical context. Robert Harbison put it succinctly: "Catalogues make their users authorities on subjects they have not heard of till that moment, abolish the beginning and end of learning."[34] Somehow the heft of the catalog conveys authority on its use, even its misuse. At a moment when engineering and architectural training were diverging, catalogs afforded unfiltered and quasi-authoritative access to the formal products of science.

The Geometry of Bricolage

As architectural and engineering education diverged in the early twentieth century, architects were no longer exposed directly to mathematical geometry as a matter of conventional training. The influence on design of the attendant scientific maquettes and classificational catalogs began to subtly mutate by the mid-twentieth century. On the one hand, some designers saw mathematical models as specific formal archetypes that could define a space independent of their methodological origin, akin to Ernst's collaged assemblages. Le Corbusier (1887–1965), in particular, seemed to align with this tendency. On the other hand, certain designers took models as methodological exemplars, archetypes of an underlying procedural design technique that could be expansive and adaptive, yet deductively precise. Apart from the charms of any single model, such techniques offered new ways of seeing, thinking, and making.

Le Corbusier embraced maquettes as objets trouvés, suitable for direct insertion into the architectural lexicon on their aesthetic merits alone. In his books, including *Vers une architecture*, he treated all manner of industrial products as formal fodder for design, and his approach to mathematical objects extended this practice. His projects integrated, almost bodily, volumes influenced by mathematical maquettes. Two projects in particular had indispensable connections to geometric model culture: the Phillips Pavilion and the Church of Saint-Pierre. Designed for Expo 58 in Brussels, the Phillips Pavilion housed the *Poème électronique*, a light and sound multimedia performance experience enabled by Phillips electronic equipment.[35] Composed of a sequence of hyperbolic paraboloid and conoid patches, the Pavilion was a ruled-surface project par excellence. Le Corbusier's collaborator Iannis Xenakis (1922–2001) primarily developed the ruled-surface hypars and conoids that enveloped the project. Yet, even in their early research for the project, Le Corbusier and Xenakis had gathered manuals and catalogs featuring ruled mathematical models. Marc Treib noted that Le Corbusier contacted the University of Zurich and the ETH Zurich asking for what he called a "catalog of forms": mathematical functions from which he might derive an architecture suitable for the pavilion.[36] The director of the Institute of Mathematics at the University of Zurich, B. L. van Der Waerden (1903–1996), suggested Jahnke and Emde's *Funktionentafeln mit Formeln und Kurven*, the same manual from which Ernst had drawn less than a decade earlier. Le Corbusier also retained in his library a catalog of mathematical models from the Brill-Schilling factory in Halle, Germany, which showcased wire models similar to Olivier's. The extent of their direct sampling of these models is impossible to determine, but Le Corbusier and Xenakis were clearly aware of manuals and catalogs of mathematical geometry that could be readily used as sourcebooks for form.

figs. 3.12, 13   The unselfconscious use of the vocabulary of ruled string models is perhaps even more striking in Le Corbusier's Church of Saint-Pierre, in Firminy, France, designed after Xenakis's departure from the office. Developed primarily between 1961 and 1963, the composition of Saint-Pierre is far simpler than that of the Phillips Pavilion, consisting simply of an ellipse traced on a ramp plane cutting through an invisible vertical cylinder, and subsequently connected by ruling lines to a square base.[37] It is impossible to ignore the strong affinity between the study models of this project and Olivier's string models of a century before, which appeared in the mathematical catalogs that Le Corbusier had collected. Even the early schematic sketches of the project delineate the ruled surfaces in a clearly identifiable way. Ruled-surface string models seemed to provide Le Corbusier with a ready stock of forms at once strangely alien yet identifiably architectural.

Methods of Calculation and Intuition

If Le Corbusier played the bricoleur, the brothers Naum Pevsner (better known as Naum Gabo, 1890–1977) and Antoine Pevsner (1884–1962) drew more sculptural and intuitive effects from mathematical form. Virtuosos of abstract shape, they carved volume from air itself, bounded merely by line or paper-thin surface. Yet, despite their similarities, a close reading of their mathematical approaches suggests divergent sensibilities aligned with distinct philosophies of design. On the one hand, one can trace a clear Anglophone line of descent from Naum Gabo's interplay between mathematical precision and design intuition to the Cambridge University architecture department of the 1960s, the Design Methods movement, and procedurally inclined architects like Lionel March, Christopher Alexander, and Peter Eisenman. Antoine Pevsner's heirs, on the other hand, included Max Bill, Germany's Ulm School of Design of the 1960s, and a distinctly Teutonic sensibility in computational design. Over time, these parallel lineages adopted distinct ideologies of rigorous design with echoes that still reverberate.

In the MoMA catalog for their 1948 retrospective, curator Herbert Read argued that scientific notions of space were always entailed in Gabo and Pevsner's work:

> The particular vision of reality common to the Constructivism of Pevsner and Gabo ... is derived not from superficial aspects of mechanized civilization, nor from a reduction of visual data to their "cubic planes" or "plastic volumes," but from the structure of the physical universe as revealed by modern science. The best preparation for a true appreciation of constructive art is a study of Whitehead or Schrodinger.[38]

3.12   Le Corbusier, tension string study model for the shell of Church of Saint-Pierre at Firminy.
Source: Robin Evans Slide Archive, Architectural Association Archives, London.
© F.L.C. / ARS 2020 and José Oubrerie. Conception: Le Corbusier, architect;
José Oubrerie, assistant (1960–1965). Realization: José Oubrerie, architect (1970–2006).

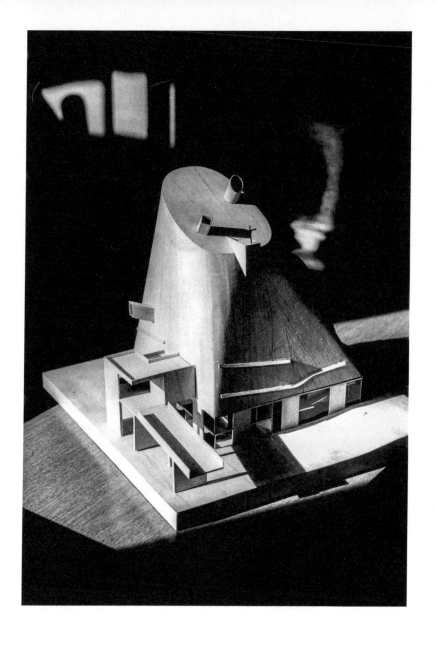

3.13   Model of Le Corbusier's Church of Saint-Pierre at Firminy, 1970.
© f.l.c. / ars 2020 and José Oubrerie. Conception: Le Corbusier, architect;
José Oubrerie, assistant (1960–1965). Realization: José Oubrerie, architect (1970–2006).

Gabo's training as an engineer nurtured this scientific affinity and, crucially, his intimate familiarity with mathematical maquettes. After dabbling in medicine, Gabo began studies at the Technische Hochschule in Munich (1912–1914), the same school where Felix Klein and Alexander Brill had taught decades before, and in which they had established a quintessential library of mathematical models.[39] Klein and Brill's prodigious model archive was a teaching resource and would have been inescapable to anyone studying mathematics at Munich. Ruth Olsen and Abraham Chanin observed that "in physics Gabo learned to make three-dimensional models illustrating the exact measurements of mathematical formulas."[40] More than any other place in the world, Munich was the ideal context for Gabo to learn the craft of mathematical model construction.

fig. 3.14

The formal impulses of Olivier's string models are legible in Gabo's sculptural work, which ranges from the scale of small models to building-sized sculptures. The pieces themselves are sheets of continuous strands carving lozenge-like negative spaces, held taut by rigid perimeter frames. At the small scale, these played out convincingly in pieces like *Spherical Construction* of 1937 and in his *Linear Construction* series, which spanned from the 1940s to the 1960s. The same impulses are brought astonishingly to the architectural scale with his 1957 *Project for Bijenkorf* in Rotterdam, an eight-story-tall vision of ruling lines and framing curves. Gabo's attraction to mathematics did not stop with forms but extended to language: Martin Hammer and Christina Lodder have pointed to Gabo's self-conscious use of mathematical vocabulary, for instance in his claim to be the first to rigorously apply stereometry—that is, three-dimensional calculations—to his constructions.[41] Even his documentation methods, such as his stereophotographic reproductions, mirrored those of mathematical peers.[42] And he acknowledged that he "did not invent the theory of 'developing surfaces,'" though his work was deeply influenced by those concepts.[43]

The interplay between mathematical technique and intuition was Gabo's field of creative invention. He saw instrumentation as a method of yoking technique and intuition, and in that sense, mathematical models became one of an arsenal of exact visual instruments. In his 1956 essay "Art and Science," Gabo exulted in the almost prosthetic implications of new visual techniques: "Now

3.14   Naum Gabo, *Linear Construction 04*, 1962.
© Detroit Institute of Arts, USA / Gift of W. Hawkins Ferry / Bridgeman Images.

the sciences have enabled us to grow supplementary organs to our five senses in the form of new multiple and complicated devices; we behold an entirely new image of these mysterious forces.... The artist of today cannot possibly escape the impact science is making on the whole mentality of the human race."[44]

fig. 3.15   Architects embraced Gabo's work not solely as sculpture but as proto-architectural design research. Gabo's protégé, the British architect Leslie Martin (1908–2000), framed Gabo's work as "possible prototypes for building ideas."[45] With Gabo, he affirmed the need for deeper modes of vision: "In science, as in art, 'appearance' has been jettisoned in favour of a world discovered only through the penetration of appearances."[46] Martin was vitally interested in design research in all of its varieties, particularly Gabo's constructivism. Martin's own inclinations toward constructive methods contributed to his founding of the Center for Land Use and Built Form Studies at Cambridge University in the 1960s, a research center that foregrounded exact calculational methods. Gabo's particular mathematical methodology was also admired by architects in America, and in 1953 he was invited to teach a course on "design research" at Harvard's Graduate School of Design.

While Gabo mastered the delicate transparency of Olivier's wire models, his brother Antoine Pevsner (1886–1962) worked in continuously curving sheets of brass and bronze. Like Gabo, his titles and language often paid homage to the ruled and developable surfaces which were their evident technical foundation: his *Developable Surface* of 1939, *Developable Surface* of 1941, and *Developable Column* of 1942 make this association unmistakably explicit. Graphically, these sculptures often bore the markings of their mathematical counterparts, the relevant ruling lines drawn directly on the sculptures. Yet, despite obvious mathematical references, Pevsner's coy public position on purely scientific methods was evasive: "Some people believe that I employ formal geometry and mathematics to design my sculptural spatial constructions but they are wrong.... Of course, I permit no formal geometrical calculations in designing the sculptures.... Yet, formal geometry could be considered to be poetry."[47] Evidently, the will to authorship yet overrode a complete embrace of calculation.

Still, Pevsner's two closest interlocutors, the De Stijl painter and sculptor Georges Vantongerloo (1886–1965) and the architect

3.15   Naum Gabo, *Project for Bijenkorf*, Rotterdam, Netherlands.
© Spaarnestad Photo, Nationaal Archief, Anefo Collection, 1957 /
Image: Herbert Behrens. All rights reserved.

Max Bill, had few qualms with explicit mathematical references. Like Pevsner and Gabo, Vantongerloo invoked mathematical titles, for instance the 1930 painting *Composition émanante de l'équation* $y = -ax^2 + bx + 18$. Formulas—cryptic encodings of form, like the Linnaean names of strange faunal specimens—became a mode of registering mathematical sensibility.

Twenty years their junior, Max Bill sympathized with Pevsner and Vantongerloo's penchant for mathematical forms and language. In effect Bill became the heir apparent to both constructivism (through Pevsner) and De Stijl (through Vantongerloo), fused through Bill's reformulation of them as Concrete Art. The three artists exhibited together at the joint show "Pevsner, Vantongerloo, Bill" at the Kunsthaus Zurich in 1949. The catalog for this show was the occasion of Bill's pivotal essay "Die mathematische Denkweise in der Kunst unserer Zeit" ("The Mathematical Approach in Contemporary Art"), in which he explicitly called for new visual methods tied to the lineage of mathematical models: [48]

> I am convinced it is possible to evolve a new form of art in which the artist's work could be founded to quite a substantial degree on a mathematical line of approach to its content ... like those [plaster mathematical] models at the Musée Poincaré in Paris where conceptions of space have been embodied in plastic shapes ... they undoubtedly provoke an aesthetic reaction in the beholder.[49]

In these models, Bill saw beyond the superficial geometric novelty of exotic surfaces to a general system of topological organization. He was one of the first architects to seriously take on topological experiments, a topic which will receive greater attention in chapter 7. Bill would become one of the most significant European design educators of the twentieth century through his work as the rector of the famed Ulm HfG design school. But his explicit theoretical embrace of geometric maquettes as a methodological precedent linked the visual language of nineteenth-century mathematics and the advent of topological impulses in design. Through topological thinking, mathematical geometry could grow from the maquette scale into a bona fide paradigm of architectural design.

The Architectural Uses of Mathematical Things

As modern architects hunted for design's future in an increasingly technical culture, some found clues in mathematical models, which became muses, archetypes, and bridges to a new universe of calculated figuration. Intricate maquettes acted as a cultural currency exchanged and circulated between mathematics and design, diffusing not only tangible forms but sensibilities of representation. They comprised an enticing lending library of self-evidently rigorous figures, extracted, adapted, and redeployed in the service of design. Moreover, mathematical models became tokens of initiation, orientation, and desire, common references of emerging social clusters persuaded of the power of exact methods in architecture. Mathematical models sowed the seeds of a thousand formalisms.

Mathematical maquettes confronted architects with processes of form making that were not isomorphic to their products. In mathematical models, process and product operated on different planes, and through different languages. Until the twentieth century, isomorphism was the entire raison d'être of architectural drawing: the fidelity of drawing to building ensured that the drawing itself was an experimental proxy through which the building could be iteratively interrogated. Mathematical models were a different proposition: they were outputs of processes that bore no visual resemblance to their products. Like a shadow play, they dislocated the process of design from a direct crafting of the model or drawing itself to an indirect manipulation of its encoded form.

The inclination of designers to engage robustly with the underlying mathematics of these models fell along a broad spectrum. For some, like Le Corbusier, the generating methods were almost beside the point if the artifacts themselves were amenable to architectural adaptation. Others were perhaps perversely attracted to the inscrutability of mathematical form: the models were readouts of an arcane science, telegraphed signals from another reality, enigmas to be deciphered or interpreted as much as tools to be used. For yet others, like Leslie Martin, mathematical methods were far more consequential than the forms recorded in models, and new training to equip architects with mathematical fluency could unlock new ways of conceiving design itself. This last impulse led to a decades-long drive to mold architectural intuition to a calculational mindset. To put it simply, it seemed that to advance architecture, designers must become geometers.

Regardless of their literacy in the underlying techniques, as designers adopted the formal accoutrements of mathematical models, they absorbed certain cultural and methodological conventions and sensibilities. Models were inscribed with the markings of the calculation process, which designers emulated in their own models. String models retained ruling lines, and plaster models were marked with characteristic curves and inflection points. Vestigial graphics, like sacral markings, persisted in mathematically oriented design maquettes. Models showed what rigor looked like as product and process, and thus became didactic props in the creation of a social cohort of mathematically inclined architects. Linguistic cultural markers were also imported, as technical geometric vocabulary began to filter into architectural conversations. Designers strove to speak more exactly and credibly of the forms they so admired, to furnish architecture with a polemic language equal to the radical new visual language of these models. The result was the evolution of a geometric lexicon specific to twentieth-century design culture, stitching together the language, forms, and values of both mathematics and architecture.

Designers intrigued by the convergence of architecture and mathematics had to confront the contradictions latent in the epistemology of architecture. As in Max Ernst's collage, mathematical fact collided uneasily with creative archetype. Were general principles of knowledge possible when individual creative vision was at the center of architectural authorship? Mathematical design impulses challenged architecture's preoccupation with the lone author, begging the question of its status as contingent art or deductive science.

In 1950 the pseudonymous mathematician Nicholas Bourbaki famously called the epistemology of deductive theorems the "architecture of mathematics."[50] Bourbaki's premise was that the edifice of mathematics rested on a handful of common, universal, and logical axioms that were sufficient to deductively derive the vast galaxy of theorems that comprised even its most arcane subdisciplines. Such a philosophically rigorous notion of knowledge presented a daunting ideal for certain designers. In the contingent space of architecture, necessary or tautological premises seemed ill suited or even impossible. Architecture was arguably conditioned as much by individual judgments of taste or specific situations as by general and objective knowledge. The tensions between knowledge and desire demand that architects make formal choices for which there may be little deductive rationale. Yet design is still bounded by stubborn facts, systems, and external agencies, all of which have objective dimensions. The attraction of mathematical models was that they hinted at a third way: a vocabulary that followed from logical necessity but furnished the limitless formal invention that design demanded. Scientific models seemed to be the antidote to the subjective relativity and ambiguity of design. In their transcendent embodiments, mathematical maquettes were guideposts toward a rationality that could catalyze new extremes of creation.

4   Alternate Dimensions: Measuring Space from Building to Hyperbody

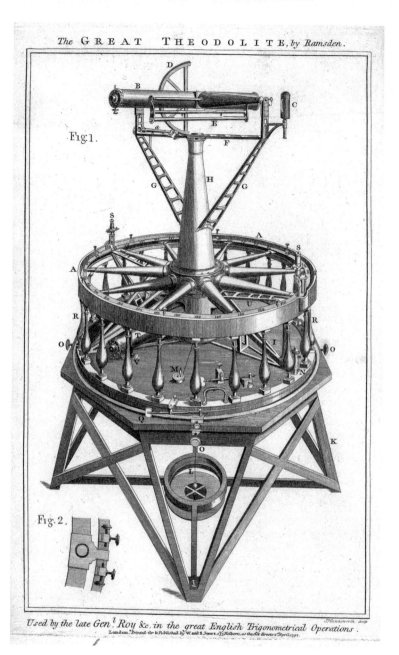

The GREAT THEODOLITE, *by Ramsden.*

Fig. 1.

Fig. 2.

Used by the late Gen. Roy &c. *in the great English Trigonometrical Operations.*

London, Printed for & Published by W. and S. Jones, N.º Holborn, as the Act directs, April 1797.

4.1   Jesse Ramsden's theodolite, the most influential iteration of the instrument in its modern form. Source: William Roy, "An Account of the Trigonometrical Operation, Whereby the Distance between the Meridians of the Royal Observatories of Greenwich and Paris Has Been Determined. By Major-General William Roy," *Philosophical Transactions of the Royal Society of London* 80 (1790), pl. III.

fig. 4.1

Throughout the eighteenth and nineteenth centuries, optical instruments surveyed buildings and land with ever-intensifying precision, allowing the complex geometry of physical space to be scanned into simple sequences of numeric measurements. Surveying was an exact visual accounting: through the meticulous tabulation of point locations, visible space was reformatted as an abstract ledger of gauged dimensions. For the architect, surveying allowed the heavy forms of buildings and landscapes to be translated into the weightless realm of mathematical entities. As Robin Evans explains, "While architecture provided the opportunity to force substance into geometric shapes, the allied business of surveying provided an opportunity to divest geometry of this same task as soon as it was turned from the designing of perfect architectural forms to the recording of imperfect ones."[1] Surveying foregrounds rational intermediaries we place between ourselves and the world, but also calls attention to that in the world which we deem irrational. Surveying is a rectification and rationalization of irrational form.

Architectural historian Mario Carpo has pointed out that surveying is, and always has been, an essential mode of computational representation.[2] It is a practice that erects graphic and numeric proxies for reality. Those proxies, from triangulated maps to abstract multivariable parametric functions, were catalysts for a corresponding expansion in architecture's modes of drawing and encoding space. Gradually, architectural representation grew to encompass not only spatial measurements but innumerable other parameters of activity, woven into the fabric of mathematical networks of points and lines known as triangulated meshes. The triangulated mesh was a specific type of survey drawing, but it was also a manifestation of collective regulation: regulation of Earth, of architecture, and of the myriad seen and unseen forces that shape them. It mapped and embedded architecture in a wider matrix of geographic, political, and natural logics, tying them together in a common and mathematically rational framework.

The triangulated meshes produced by mathematized surveying are quintessentially virtual representations, duplicates in data of real or imagined spaces. They are reformatted realities, forms recorded in a self-referential pattern of delicate recursive subdivision. As a mathematical construct intended as a proxy for physical things, the mesh abstracts material and materializes abstraction. As diagrams, mathematical mesh representation suppresses the tactile and phenomenal properties of material— texture, volume, mass—and replaces these with pure spatial information. The mesh remakes all objects representable by triangulation on the same ontological level.

Its conceptual versatility allows the triangulated mesh to act as an adaptable device that reveals the preoccupations of the times that employed it. In effect, the mesh is a cultural document that demarcates not only space but rationality. In fractious seventeenth-century Europe, meshes optimized the design of fortified cities and provided a graphic framework for maneuvers within militarized landscapes. In the nineteenth century, meshes delimited commercial networks or documented land for colonial exploitation.[3] Meshes later became the geometric shorthand for structural systems of curved shapes in the twentieth century, such as Buckminster Fuller's archetypal dome-meshes.[4] Today, meshes are used in digital tools to represent and quantify a range of environmental, ecological, and operational controls such as solar analysis, wind simulation, structural optimization, or waterflow visualization. As abstract structures, meshes require an orientation that invests them with specific meaning. They are devices with evolving cultural biographies.

With German mathematician Bernhard Riemann's 1854 discovery of geometries that exceeded three dimensions, the surveyed drawing could no longer draw all of reality. The limits of rationality itself irrevocably expanded. New measures of architecture such as time, cost, or energy could be adjoined to length, width, and height. These new dimensions parameterized space as a multivariable function. As design became accountable to new parameters of measurement, symbolic encodings of space emerged based on a multivariate parameterization that exceeded the purely graphic representations of typical architectural or surveyed drawings. By promising designers a way to answer multiple objectives as never before, multivariable functions also dethroned drawing as the sole valid architectural representation.

The pas de deux between these two representations—the surveyed map and the multivariable function—betrays a more general tension between graphic drawing and symbolic encoding.

These tensions in turn reveal the central but precarious place of optical vision in architecture's evolving representations. Surveying was hypertrophied vision, the eye ranging across Earth to bound and demarcate it. Survey instruments operated through the natural faculties and constraints of human sight, its rules of perspective and its contained horizon. In contemporary design, triangulated meshes encompass whole sets of overlapping but unique data: maps of curvature, material stresses, and environmental performance. Yet the multiplication of dimensions bearing on architecture—dimensions not only of space but of performance, quality, and experience—implied that some of these might be invisible, and that the eye could no longer hold architectural form in complete thrall. New practices of augmented vision and mathematical drawing could expose these unnatural new dimensions and correlate them with the natural ones, but new training and distinct intuitions had to accompany new visual capacities. Symbolic encodings and mathematical functions became the native language of these more complex realities. The unaided eye no longer ruled architecture, while mechanical modes of augmented vision and hyperdimensional forms of mathematical drawing encompassed not only architecture but the city, the landscape, and Earth itself.

Triangulation as Formal Encoding

Surveying is an essential way to reconcile the irregularity of Earth to the regularity of architecture through a common regime of rational vision. Surveying sees all surveyed objects on equal terms. The centrality of human sight in classical surveying practice tied it inextricably to the related practice of surveillance, to which it was also etymologically related. Surveying was surveillance multiplied. If surveillance was a single observer intensely attending to a particular space, surveying coordinated surveilled spaces into a collective consensus of observation. As vision moved from the individual to the collective, it also shifted from subjectivity to objectivity. The recording of collective observation was a practice that was both graphic and tabular, drawn and numeric.

Pragmatically, surveying not only determines the topography of landscape but also assesses the condition of buildings within it, from simple elevation levels to more detailed measured plans. For centuries, the devices to measure land and the devices to measure architecture—levels, metric chains, and compasses—were identical, and thus surveying and its attendant virtualizations of space constantly informed the imagination of architecture itself. As its instruments and practices grew more exact, surveying gave rise to peculiar new forms of representation which harmonized with an increasingly quantitative view of space. The triangulated map was a particular genre of mathematical drawing which, more than any other, epitomized a calculated and rational way to delineate irregular landscapes.

The technique of triangulation was first developed in sixteenth- and early seventeenth-century Holland to reckon the size and shape of the entire Earth. Originally proposed by the mathematician Gemma Frisius (1508–1555), it was later applied by the cartographer Willebrord Snel van Royen (known as Snellius, 1580–1626) to measure the true geodesic distance between the two small Dutch towns of Alkmaar and Bergen op Zoom.[5] After taking a series of angular measures and estimated distances between the steeples of local churches, Snellius balanced a set of trigonometric equations to converge toward an extraordinarily accurate estimate of the distance between the two towns. From this distance he extrapolated a startlingly close estimate of Earth's entire circumference. The accuracy of this method was a dramatic improvement over prior methods, and triangulation soon proved useful for registering complex forms of all types, including cities, parcels, landforms, or buildings. Triangulated surveys released space from the bounds of the Cartesian grid and shifted it into the language of relational geometry.

From the moment of their inception in the sixteenth century, triangulated meshes were exact tabulations of inexact forms. All of the irregularities of geology or landform could be recorded and rationalized precisely. They virtualized natural Earth and artificial architecture in a common, abstract, multiscalar web of visible and measured relationships. The ontological leveling process of triangulation treated every space and object, no matter how regular or irregular, as amenable to the same technique of scanning, recording, and mathematical drawing.

Following Snellius's example, later surveyors located clusters of dozens or hundreds of observation nodes, with distances between them estimated by special surveying chains laid between nodes. The resultant triangulated maps allowed cartographers to create detailed surveys of landforms, capturing the elevations and various features in the terrain with more precision than ever before. Through laborious trigonometric calculations, a cartographer could rectify the measured lengths of the chains and converge toward a mutually coordinated network of true dimensions. The historian Matthew Edney notes that such a network was integral and self-consistent, "a rigorous mathematical framework in which all points are defined with respect to each other."[6] The triangular network was a cage for form, an envelope that did not replicate the surveyed terrain identically but contained it within an arbitrarily close proximity. By the early nineteenth century, triangulation became a standard and widely used technique for surveying throughout France, England, and western Europe generally.[7]

Methodologically, triangulated meshes replicate and multiply the apparatus of perspective across an extended range. With a plurality of perspectives, meshes tended toward what Lorraine Daston calls an "aperspectival objectivity" that "opposed the subjectivity of individual idiosyncrasies."[8] Meshes are thick webs of mutually visible perspectives, stitched together through the measuring eye of the surveyor and a mechanical-optical device called the theodolite. The theodolite was the indispensable instrument for optical surveying and triangulation. It took its modern form by 1790, when the British instrument maker Jesse Ramsden (1735–1800) explicated his version of the instrument to the Royal Society of London.[9] Though precursors had been used for triangulations since the sixteenth century, this definitive new version was a massive 200-pound instrument that consisted of a substantial wood and iron base upon which a doubly gimballed telescope was placed at a height of about five and a half feet.[10] Once the theodolite was hefted into place at an observation point, the surveyor would adjust the viewfinder and peer through it to sight pairs of distant features—typically miles away—and record the angles between them. Two orthogonal dials, one horizontal and the other vertical, controlled the rotation of the viewfinder in each plane. The dials also indexed horizontal and vertical angles, and their

fig. 4.2

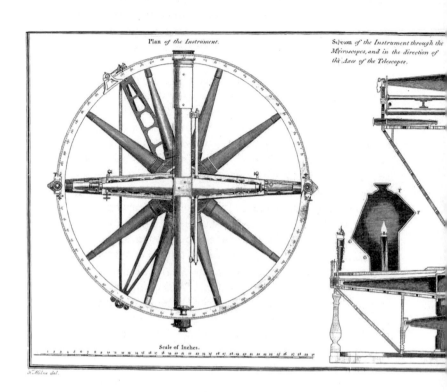

Plan *of the Instrument.*

Section *of the Instrument through the Microscopes, and in the direction of the Axes of the Telescopes.*

Scale of Inches.

4.2   Plans and sections of Ramsden's theodolite.
Source: William Roy, "An Account of the Trigonometrical Operation, Whereby the Distance between the Meridians of the Royal Observatories of Greenwich and Paris Has Been Determined. By Major-General William Roy," *Philosophical Transactions of the Royal Society of London* 80 (1790), pl. IV.

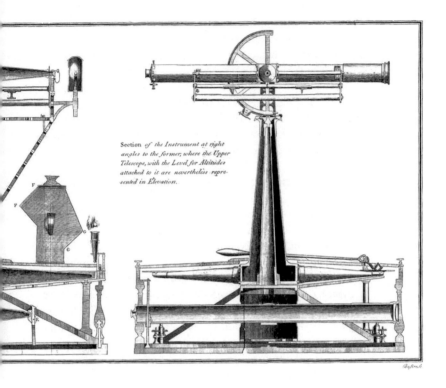

Section *of the Instrument at right angles to the former, where the Upper Telescope, with the Level for Altitudes attached to it are nevertheless represented in Elevation.*

F

F

G

G

measurements provided the relative angular distance between the survey points in a triangulation. Each remotely observed node—hilltop, steeple, roof—was a distinct perspectival position forming a network of mutually connected, codetermined, and interlaced locations. This process of sighting and angular measurement was repeated for each of the dozens or hundreds of points in the network. The detail resolution of the resulting map was limited only by the density of nodal survey points and the machinery of calculation. The result, as the historian of science D. Graham Burnett observed, was a mathematical map that was "closely congruent to the land itself."[11] The sightline connections between remote observation points folded the idealized and infinitely extended planar landscape of Renaissance perspective into an undulating, articulated, faceted terrain.

The Measured Traces of Land as Architecture

In providing a format to negotiate irregular land with regular architecture, triangulation also allowed architecture to adopt the contingency of landscape. Apart from its capacities as a medium to record and virtualize what already existed, the triangulated mesh was also a design medium for imagining geometry beyond the perfect clarity of Platonic forms. It set out the terms of negotiation between architecture and Earth. As surveying was applied toward military ends, triangulation also oscillated between a technique of political subjugation and an auxiliary of spatial imagination. Measured and rational vision proved indispensable to both.

fig. 4.4

Triangulation's rational control of an irrational world coincided harmoniously with enlightened absolutism. The surveyed control of land first intersected architecture *in extremis* in the design of city-scale military fortifications at the disputed frontiers of Louis XIV's France. The absolutist state constructed networks of military fortresses to monitor and control its land, and their planning applied the surveying devices and techniques toward the geometry of tactical warfare. The preeminent practitioner of the geometrized militarization of landscape was the seventeenth-century French military architect Sébastien Le Prestre de Vauban (1633–1707), the Maréchal de France during the reign of Louis XIV.[12] Vauban articulated a geometry and practice

110

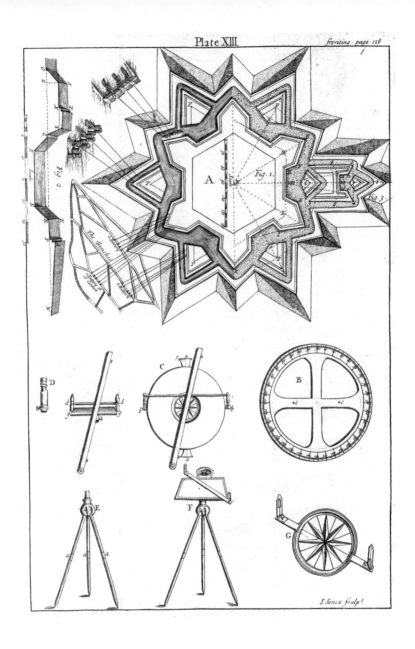

4.3   Nicolas Bion, fortification surveying tools.
Source: Nicolas Bion, *Traité de la construction et des principaux usages des instruments de mathématique*
(Paris: Jean Boudot, 1709), pl. XIII.

4.4 A plan of a proposed Vauban fortification, illustrating the triangulated subdivision of battlement walls. Source: Sébastien Le Prestre de Vauban, *De l'attaque et de la défense des places* (The Hague: Pierre De Hondt, 1737), pl. 23.

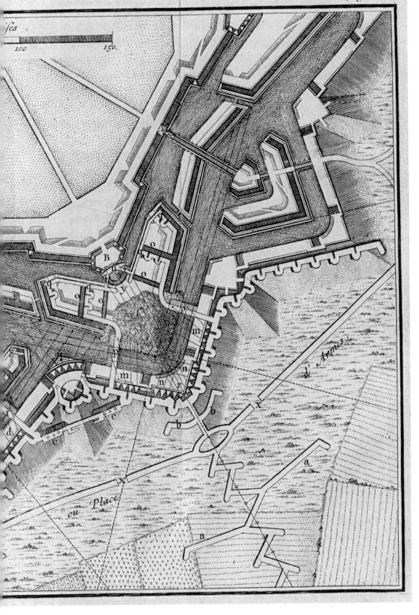

of siegecraft and fortress building that was codified posthumously in the 1737 *De l'attaque et de la défense des places*.[13] Over a series of studies and projects for bastion towns at the border between France and the Spanish Netherlands, Vauban designed starlike geometric citadels with radiating points that compromised the formations of attacking infantry forces. Vauban's most durable spatial innovations were in the design of vast fortifications with massive polygonal earthwork walls. By translating the parameters of spatial control into a mathematical system of wall angles, infantry arrangements, and ballistic arcs, Vauban geometrized state power through an architecture of surveillance and surveying.

The recursive triangulation of Vauban's fortified city walls partitioned the landscape around them into optically regulated and easily controlled segments. The exigencies of siege craft shaped not only the plans of these citadels but also their sections: the parabolic paths of ballistic projectiles determined the proper height and geometry for defensive bulwarks and ramparts. Accurate surveying was indispensable to the derivation of fortress geometry, literally a life-or-death calculation. Vauban's faceted, fractalized drawings, falling somewhere between surveyed maps and architectural graphics, were a particular type of triangulated mesh generated from a network of privileged surveillance points. The surveys of raised fortifications were calibrated not only to measure land but to partition vision and control of the landscape below. Rather than being imposed on the landscape, rational geometry defined a constantly negotiated membrane between irregular landscape and regular architecture.

fig. 4.5 The French application of triangulated surveying rapidly expanded from the scale of the city-fortress to that of the state. Historically, the measurement of land at a large scale was fraught with errors, defects, and inaccuracies, subject to the distortions and irregularities of physical measuring ropes and chains. The optical process of triangulation removed these impediments and turned large-scale surveying into an exact science. Nowhere was it embraced more fully than in seventeenth-century France, in which triangulated cartography became an integral organ of the state bureaucracy. In 1669 the French astronomer Jean-Félix Picard (1620–1682) began a triangulation of the meridian between the northern coast of France, through Paris, to the southern coast.[14]

He calculated meshes spanning the country, estimating the most accurate figure for Earth's circumference yet. Jean-Dominique Cassini (1625–1712) and three successive members of the Cassini family—by royal appointment the heads of France's national observatory—expanded Picard's already ambitious project to triangulate the space of the entire nation. They undertook nearly a century of triangulation work to complete a rigorous national cartographic survey by 1744.[15] In effect, this survey accurately measured the space of France for the first time. The centralized state demanded seeing its territory with a total, exact, and rational eye.

In the nineteenth century, the surveying eye was conscripted as a critical auxiliary to the organization of cities, counties, regions, and nations in Europe and eventually America. Triangulated maps became a graphic index of increasingly technocratic states. The execution of these surveys was often a military function, sometimes carried out as an administrative aspect of imperial conquest. When Napoleon's army advanced across continental Europe, they triangulated subjugated territories, and thereby executed some of the first triangulations of Prussian land.[16] The historian Ernst Breitenberger noted that these military exchanges diffused the technique across the continent, and triangulations of Denmark, Austria, the Netherlands, and elsewhere began soon after the conclusion of the Napoleonic wars.[17] Sprawling inventories of national and imperial space were soon inaugurated, driven by the commercial need for precise coastal charts and enabled by theodolites derived from Ramsden's design. Often taking decades to complete with an army of peripatetic surveyors and calculators, triangulations were mammoth undertakings, initiated both for pragmatic ends and as bureaucratic monuments to the modern states that sponsored and managed them.[18] Among the most expansive triangulations was the seventy-year Great Trigonometrical Survey of India begun in 1802, which, among other achievements, mapped the exact heights of the Himalayas through triangulated meshes.[19] Mount Everest, for example, was named for George Everest (1790–1866), one of the superintendents of the Trigonometric Survey.[20] With increasingly precise survey instruments, these cartographic programs generated exact virtualizations of the land and used the device of perspective projection at an immense scale to quantize national and colonial territory.

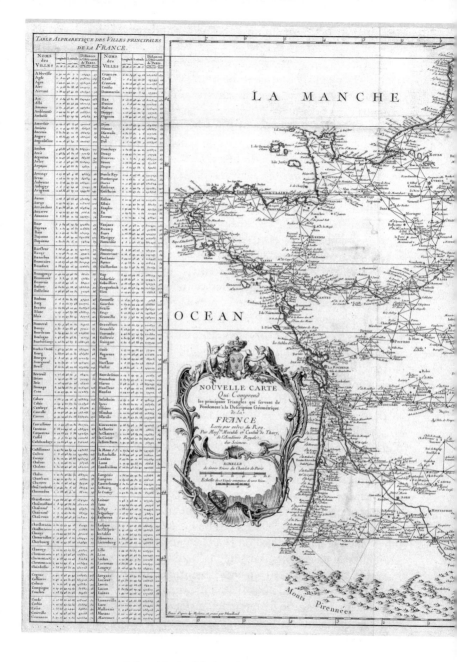

4.5   A plan of the Cassini triangulation of France, 1744.

Source: *Nouvelle carte qui comprend les principaux triangles qui servent de fondement à la Description géométrique de la France. Levée par ordre du Roy par Messrs. Maraldi et Cassini de Thury, de l'Académie royale des Sciences* (Paris, 1744). Bibliothèque nationale de France, GE BB-565 (A7,10).

Ye BB. 565a VII -10

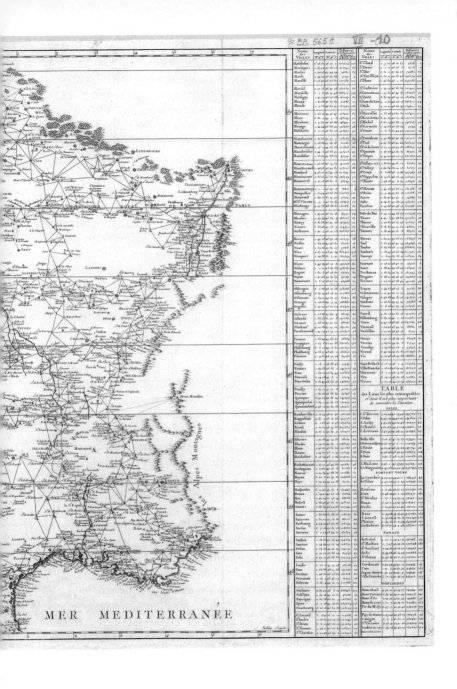

Triangulation provided a purely geometric and homogeneous vision of mathematized space at an imperial scale.

As triangulated surveying was applied at ever more vast scales, it became an increasingly mathematical endeavor. At the heart of this scientific evolution was the geometric question of the exact shape of Earth itself. By the early nineteenth century, it was clear that Earth was not a true sphere but rather a deformed spheroid. But in order to determine the exact shape, massive geodesic arcs had to be triangulated through meridians and across the equator itself. The calculations entailed in these triangulations were so exceedingly precise and so profoundly difficult that in nineteenth-century France and Germany their study emerged as a mathematical subdiscipline in its own right: geodesy. Geodesy fused cartography and advanced surface geometry into the same discipline.[21] The theory of geometry and the practice of surveying mutually reinforced each other. For instance, mathematicians developed theoretical innovations like quantitative characterizations of surface curvature, which were duly applied to large-scale mapping problems. Throughout the nineteenth century, theoretical geometers were surveyors and vice versa: the imminent German mathematician Carl Friedrich Gauss (1777–1855) and French mathematician Henri Poincaré (1854–1912) were both involved in national triangulation projects, and devoted considerable energy to geodesy and spatial surveying, which they saw as a natural corollary of their geometric investigations.[22]

Technologies of vision also transformed the practice of surveying, placing new machinic intermediaries between the human eye and the land. In part, that transformation proceeded through the evolution of theodolites themselves, as they progressed from tools to direct the unaided eye to telescopic devices. But the advent of photography provided entirely new options for encoding visual information. Though triangulations were typically calculated from manually collected sextant or theodolite readings, the advent of field photography in the 1830s furnished an alternative for previously intractable problems of land and building mensuration. Observed angles measured on site with the viewfinder of a standard mechanical theodolite could be replaced with constellations of photographs that were methodically compared and coordinated off site. The apparent parallax of identical features in distinct photographic plates allowed a draftsman to infer surveying distances without the time-consuming calibration of theodolites in the field. Angular measures could be derived in the comfort of an office, where previously distant or inaccessible features could now be measured with precision. Photogrammetry, as this new mode of surveying was known, provided a powerful new technique to record and dimension the visible world.

fig. 4.6

Triangulation and photogrammetry rendered vast planetary scales visible and therefore measurable, and the same techniques were soon adapted in microcosm to mathematically metrize individual buildings. Photogrammetry could be used to scan existing architecture as never before, in effect creating mesh triangulations of buildings that virtually recorded their many dimensions. The German architect Albrecht Meydenbauer (1834–1921), who coined the term "photogrammetry" and developed its architectural application, combined photography with methods of triangulated surveying to encode buildings for cultural preservation as early as 1867.[23] To aid his project, Meydenbauer developed a novel instrument that the historian Jörg Albertz called a "photographic camera and a measuring instrument in one single system."[24] Meydenbauer was particularly enthusiastic about the recording of churches and public spaces with few documentary records and many irregularities. To that end he meticulously applied photogrammetry to document these buildings. The historian of photography Miriam Paeslack has traced how Meydenbauer hoped "to erect a photogrammetric archive of Prussia's architectural and artistic monuments," a catalog of imagery and their associated triangulations to be housed in a new Royal Prussian Photogrammetric Institute.[25] Meydenbauer and his collaborators took over 11,000 photographs of hundreds of buildings across Germany.[26] The new Institute's imperial affiliation was no accident: just as surveying struck lines of control across far-flung territories, closer to the capital it captured the architecture of Germany's ascendant empire. Architecture and Earth were woven together in a calculated mesh through the common medium of photogrammetry.

The precise inexactness of triangulated forms allowed the natural world and artificial buildings to occupy the same continuous geometric space, bringing the whole universe of form under the same governing armature. That equivalence

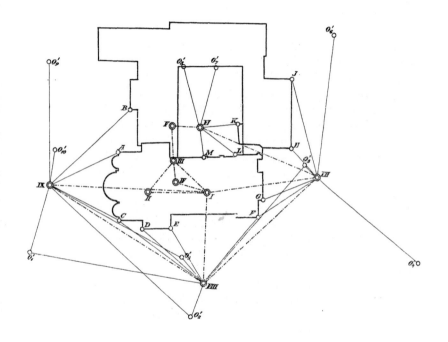

4.6   Photogrammetric surveying of Klosterkirche Jerichow (Jerichow Monastery), Germany.
Source: Friedrich Schilling, *Über die Anwendungen der darstellenden Geometrie*
(Leipzig: B. G. Teubner, 1904), 159, 163.

allowed architecture to flirt with the irregularity of natural systems, or even to adopt the geometry of the entire planet. By the mid-twentieth century, geometrical surveying was turned from triangulating the spheroid of Earth toward the triangulation of spherical buildings. In fact, for Buckminster Fuller's geodesic fig. 4.7 domes, it might be more accurate to say the triangulation of Earth and building were congruent.

For Fuller, triangulation was a framework for mapping and communication, as well as a diagram of structural robustness. On the one hand, he saw triangulation as a framework for massively amplifying the visual aspect of terrestrial surveying. He imagined a virtual globe generated by a constellation of surveillance satellites "in fixed-formation flight positions around Earth, with one such fixed satellite hovering steadily over each vertex of a one-mile edged world-triangulation grid."[27] On the other hand, triangulation was a structural principle resonant with the natural subdivision of spheres. Due to the perfect symmetry of the sphere, Fuller's domes may seem recursively symmetric, uniform to the last detail, assembled from standard modular elements. It is true that the domes were typically derived from a regular icosahedron, consisting of twenty identical triangular faces, projected onto an underlying sphere. But in fact, each of these twenty spherical triangles was itself divided into dozens of irregular subtriangles, each node a contingent measurement of a doubly curved body. The triangulated mesh proved the ideal armature to negotiate this regular irregularity. The methods for calculating this most space-age of architecture were identical to the techniques of surveying the irregular form of Earth itself deployed by the Cassini family, practiced by Vauban, and perfected by the national triangulations of the nineteenth century. The extent of the mass customization of Fuller's structures becomes clear in the drawings of his renowned us pavilion for Expo '67 in Montreal. The drawings are triangulated maps of a nonstandard architecture, dendritic networks which reveal a precise but imperfect encoding of the doubly curved sphere. In these drawings, cartographic technique and architectural representation find a strange isomorphism, a resonance of mathematical encoding and architectural form.

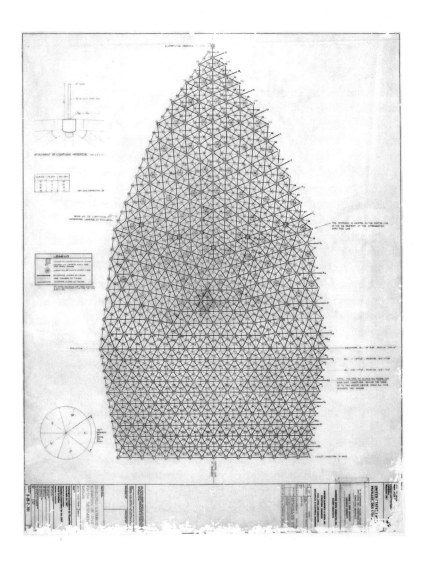

4.7 Construction drawing for Buckminster Fuller's Expo '67 United States Pavilion
showing the irregular triangulation of the constituent bar structure.
Source: The Estate of R. Buckminster Fuller. Courtesy The Estate of R. Buckminster Fuller.

Drawing Hyperspace

At the close of the Belle Époque, the eye had ascended to the peak of its influence: the total measurement of built and natural entities and the comprehensive dimensioning of reality seemed within reach through exact ocular measuring techniques like triangulation. Yet countervailing currents of scientific imaging of the fin de siècle—X-ray images of invisible spaces, photographic microscopy, and perhaps most decisively the developing notion of a fourth dimension of physical reality—revealed invisible domains and suggested a reality inaccessible to the eye alone. Seduced by the immediacy of their own sight, architects had been blindfolded to the universe of invisible forces that truly governed space. By the early twentieth century, scientific and technological shifts opened fresh lines of visual research particularly centered on the recording of hidden dimensions. While scientists learned to gauge the elusive forms and arcane terrains of four and more dimensions, architects began to imagine buildings in terms of these new dimensions and to unlock surreal new avenues in their encoded representation. The architectural theorist and historian Sigfried Giedion ascribed epochal significance to these mathematical developments:

> Around 1910, an event of decisive importance occurred: the discovery of a new space conception in the arts. Working in their studios as though in laboratories, painters and sculptors investigated the ways in which space, volumes, and materials existed for feeling.

> The speculations of the mathematical physicists seem very far removed from reality and from practical affairs, but they have led to profound alterations in the human environment. In the same way the experiments of the cubists seemed to have little significance for any kind of practice—even for architecture. Actually, however, it was just such work which gave the architects the hints they needed to master reality in their particular sphere. These discoveries offered architecture the objective means of organizing space in ways that gave form to contemporary feelings.[28]

The very intimation that drawing must expand to encompass dimensions greater than the three that ordered ocular vision was a seismic shift for architecture. The three-dimensional Cartesian grid had been baked into architectural representations since Leon Battista Alberti's introduction of constructed perspective in the fifteenth century. Even in the sciences, up until the late eighteenth century, mathematical space had always reflexively implied three dimensions. But as early as 1788, mathematicians and physicists began to adjoin additional scalar quantities to the standard three dimensions, augmenting the parameters of spatial description and hinting at expanded spaces with a potentially unlimited number of variables. In that year French mathematician Joseph-Louis Lagrange (1736–1813)[29] adjoined time to the classical three dimensions in his 1788 *Mécanique analytique*.[30] He presented time as an auxiliary dimension to describe dynamic mechanical motion, arguably the first algebraic unification of four dimensions. Spaces similarly defined by more than three dimensions became known as "hyperdimensional."

Soon the triangulated mesh itself was being extended beyond recognition by new ideas of mathematical dimensionality. Whereas Lagrange broached the possibility of global dimensional systems other than Cartesian, the German mathematician August Ferdinand Möbius (1790–1868) considered local dimensional systems emergent from the plotting of triangulated maps. In 1827 Möbius, famed for the single-faced surface Möbius strip, introduced the notion of barycentric coordinates in his pivotal paper *Der barycentrische Calcül*.[31] Barycentric coordinates provided a consistent but independent set of local coordinate systems for every distinct triangular face of a mesh, creating a relativized map of its surface. At a technical level, for each point within a triangle, the barycentric coordinates of that point are three numeric weights at each of the triangle's three vertices such that the original point is the centroid of these three weighted vertices. Barycentric coordinates revolutionized the fundamental representation of triangulated meshes for topographic, computational, and topological mappings by creating relative axis systems for each face of the mesh. Lagrange's and Möbius's work implied that the structural qualities of space were not limited to the three absolute and canonical Cartesian dimensions, but that in fact additional global or local dimensions—comparable in kind, quality, and operation to the first three—were in no way contradictory.

Even at this early stage, mathematicians sensed that higher dimensions of space entailed experiential consequences. Hyperbodies brought with them the possibility of hypersensation. A watershed moment in the revelation of hyperdimensional space was Bernhard Riemann's delivery of his habilitation lecture "Über die Hypothesen die der Geometrie zu Grunde liegen" (On the hypotheses which lie at the foundation of geometry) in 1854. In it, Riemann suggested that alternative but consistent formulations of space could diverge from our typical phenomenological experience of it.[32] Space was described for the first time in excess of our powers of biological perception. While Riemann did not specifically invoke architecture, the implications were profound: additional dimensions of space promised a radical destabilization of the conventions of experience and representation. It was only a matter of time before architecture would be forced to confront such spatial puzzles.

News of a hyperdimensional mathematics remained confined to esoteric scientific circles until certain popular accounts ignited the public imagination. For architects, two French popularizers proved especially influential: the mathematician and physicist Henri Poincaré and the visual experimentalist Esprit Jouffret (1837–1904). The polymathic Poincaré made fundamental discoveries in branches of mathematics which ultimately shaped the theoretical understanding of space, including foundational work in topology and the qualitative theory of differential equations. Poincaré was born in 1854 in the eastern city of Nancy and attended the École Polytechnique in Paris, studying under the eminent mathematician Charles Hermite (1822–1901). From his education at the Polytechnique, Poincaré was familiar with the work of Monge and Lagrange and was also trained in the practice of geodesic surveying. So it should come as little surprise that Poincaré was also deeply connected with the work of France's Bureau des Longitudes, the organization tasked with extensive triangulated surveying and which he served as president for a time.[33] Like many elite mathematicians of his day, he was invested in questions of surveying and cartography: he lectured on geodesy and served as secretary for the ambitious work of triangulating the equator to definitively measure the true shape of Earth, whether spherical, oval, or ellipsoidal.[34] Poincaré had an enormous impact on promoting the notion of the fourth dimension and the general epistemology of mathematics through a series of three books: *Science et hypothèse* (*Science and Hypothesis*, 1902), *La valeur de la science* (*The Value of Science*, 1905), and *Science et méthode* (*Science and Method*, 1908).[35] These books, and particularly their excerpts in architectural periodicals, proved decisively provocative to architects of early modernism.

Though the unaided eye was no longer the comprehensive guide it once was to reality, mathematical drawing mutated to make fragments of hyperdimensional space visible. To this end, the French mathematician Jouffret's 1903 *Traité élémentaire de géométrie à quatre dimensions, et introduction à la géométrie à n dimensions* (Elementary treatise on geometry of four dimensions, and an introduction to the geometry of *n* dimensions) offered a drafting manual for higher geometry. Jouffret applied Gaspard Monge's graphic techniques of descriptive geometry to hyperbodies, the geometric figures that inhabit spaces beyond three dimensions. As if underlining the cartographic interest in hyperdimensionality, the *Traité* was published by France's Bureau des Longitudes, the official government office charged with astronomic and terrestrial surveying during Poincaré's time as a member of that august body. Jouffret, also a graduate of the École Polytechnique, presented the potential implications of four-dimensional drawing in a strikingly visual way. Opening with a reference to Poincaré's observations on dimensional perception, Jouffret offered a method of projecting hyperbodies and hypersurfaces—those described with four or more dimensions—into a mapped space of two dimensions.[36] His aim was to impart a visual intuition for hyperbodies—as he terms it, a "blindfold-play" analogous to the powers of visualization of chess masters.[37] He elaborated ways to understand coincidence, intersection, parallelism, perpendicularity, and the measurement of angles in this strange new world. Jouffret's fantastically intricate drawings of hyperdimensional polyhedra were beautiful diagrams in their own right, but they also betrayed a genetic connection to much earlier cartographic methods. The drawings are unmistakably the product of a surveying process: they enumerate every vertex, and each relationship is calculated from an unseen hyperbody. Perhaps most strikingly, Jouffret established a process for unfolding these complex figures into the plane, including the construction of intricate tabbed paper models. These models allowed invisible hyperbodies to gain a fashion of physical presence.

figs. 4.8, 9

Jouffret was one in a coterie of authors constructing mathematical drawings of hyperdimensional objects, parlaying cues from projective geometry and geodesy into visual practices at the frontiers of perception and reality. In fact, though dimensions beyond the Cartesian three seemed to frustrate any intuition, various types of mathematical drawing actually proved remarkably adept at visualizing them. Other mathematicians, engineers, scientists, and teachers—many of whom Jouffret himself cited—rushed to speculate on how higher-dimensional objects might be drafted. In some cases, like Jouffret, they employed practices of descriptive geometry in novel ways to literally unfold hyperdimensional objects onto the two-dimensional page. As historian Linda Dalrymple Henderson has noted, the American mathematician Washington Irving Stringham (1847–1909) developed many startling drawings in this manner. Stringham's 1880 paper "Regular Figures in n-Dimensional Space,"[38] a development of his PhD thesis from Johns Hopkins, described characteristics and derivations of n-dimensional polyhedra with faces of regular polygons.[39] When cut into $(n–1)$-dimensional sections and projected into lower dimensions, these generalizations of the five canonical Platonic solids also reveal mesmerizing symmetries, hinting at the mysterious beauty awaiting the intrepid explorers of higher dimensions.

fig. 4.10

Hyperdimensional Design and Its Discontents

It was in this contested nexus of the visible and the invisible, the mathematical and the embodied that architecture began to grapple with the fourth dimension in earnest around 1915. Popular accounts of mathematical and physical hyperdimensionality, such as Poincaré's and Jouffret's, remixed the terms of engagement between space, mathematics, and drawing. Hyperdimensional mathematics held the promise that invisible new quantities and measures could be brought into conversation with design and

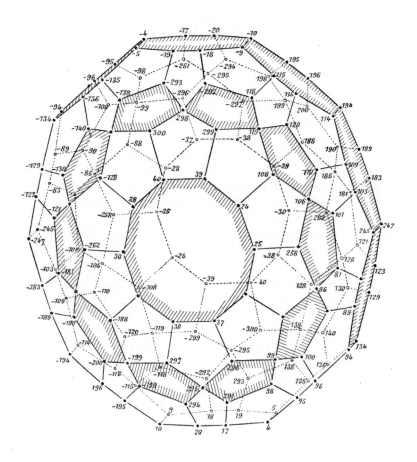

4.8 Projection of a hecatonicosahedroid, a higher-dimensional polyhedron, according to
Esprit Jouffret.   Source: Esprit Jouffret, *Traité élémentaire de géométrie à quatre dimensions, et
introduction à la géométrie à n dimensions* (Paris: Gauthier-Villars, 1903), 176.

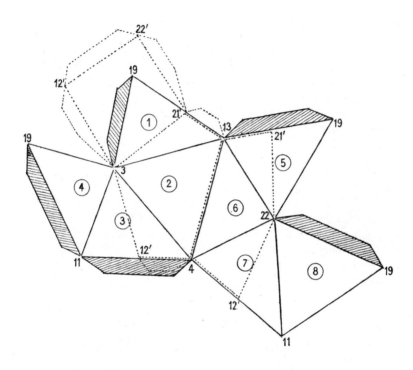

4.9   An example of Jouffret's development drawings of tabbed faceted segments that were used to construct models of larger surfaces.   Source: Esprit Jouffret, *Traité élémentaire de géométrie à quatre dimensions, et introduction à la géométrie à n dimensions* (Paris: Gauthier-Villars, 1903), 167.

4.10 Washington Irving Stringham's diagrams of unfolded hyperdimensional polyhedra.
Source: William Irving Stringham, "Regular Figures in *n*-Dimensional Space,"
American Journal of Mathematics 3, no. 1 (March 1880): 15.

bundled into more complex but holistic descriptions of space. The very relevance of the architectural drawing was called into question: if a two-dimensional drawing could only partially describe three-dimensional form, it was even less adequate to the task of hyperdimensional representation. Representation and parameterization—the process of describing shapes with mathematical functions—became inseparably linked, if not fully identical, extending techniques of drawing past recognition but also leaping beyond drawing itself to dense multivariate functions with correlated variables. Nongraphic symbolic encodings in mathematical functions seemed ascendant, as new realities required new representations equal to those realities.

If new dimensions shook confidence in old systems of both drawing and measurement, in some ways architecture was a salutatory discipline to the possibility of higher-dimensional representation. Architecture was subject to a range of invisible forces of dynamic operation and evaluation that were not well represented by drawings alone. Hyperspace offered a unique way to address and manipulate them. Even the artist Marcel Duchamp (1887–1968) recognized an affinity between architectural intuition and the promise of hyperdimensional form:

> These three-dimensional "sections" of the four-dimensional continuum . . . could be considered analogous to the two-dimensional architectural "plans" of a building. By analogy with the method by which architects depict the plan of each story of a house, a four-dimensional figure can be represented by three-dimensional sections. These different stories will be bound to one another by the fourth dimension.[40]

Before the new world of a parameterized, hyperdimensional architecture was played out in specific proposals, it was the subject of intense polemic debate in the periodicals of the day. In the public at large and in architecture in particular, the dislocating prospect of a fourth dimension of experience elicited evangelical enthusiasm from some and robust skepticism from others. The debate began early and continued for decades. In his August 1882 article for *Popular Science*, Alfred Lane discussed fourth-dimensional solids of the type Stringham had drawn, the nascent knot theory of self-intersecting curve projections, the curvature of surfaces and space, and non-Euclidean geometries. But he concludes his survey of this "transcendental geometry" with a note of skepticism: "Thus transcendental geometry, with its egg-shells turned inside-out without cracking, its knots mysteriously untied, its worlds where the background of everything is a man's own head, is from its conceptual basis, as a creation of man's mind, true. It is a pretty mathematical diversion; it is, as yet, nothing more."[41]

Some designers also cast a skeptical eye. In their essay "After Cubism—Rejection of the Fourth Dimension," Amédée Ozenfant and Le Corbusier betray considerable ambivalence about hyperdimensionality:

> The objection which here is aimed at the fourth dimension, if one wishes to reflect upon it, strikes only the gratuitous hypotheses of the theoreticians of cubism; this hypothesis is outside all plastic reality ... the fourth dimension of the mathematicians is an entirely speculative abstraction, a part of hypothetical geometry, a marvelous play of the intellect, with no material contact with the real world, conceivable but not representable, since the human senses distinguish only three dimensions in space.[42]

Implicitly, Le Corbusier and Ozenfant's argument is about the nature of reality: since the senses, including sight, survey a reality of three dimensions, invisible hyperbodies are ipso facto outside the bounds of reality. The constructivist Antoine Pevsner is yet more pointed on the limitations of an architectural fourth dimension. One can almost hear the exasperated sighs:

> Some architects delude themselves that they construct architecture in three and in four dimensions when they imitate forms taken from sculptural spatial constructions. Others introduce in their architecture geometric curves copied from analytic geometry. The third and fourth dimensions occurring in architecture are poetic creations.[43]

Indeed, early architectural engagement with the fourth dimension may have been little more than poetry. Particularly in the fever-pitch debates of the 1920s and 30s, for all the architectural discussion of hyperdimensionality, its concrete manifestation in a project seemed a distant prospect. Yet an idea, or perhaps even a vague but irresistible desire, had been irrevocably planted.

An uncharitable observer might agree with Duchamp's view that "the fourth dimension became a thing you talked about, without knowing what it meant."[44] Dilettantism was rife, and visual intuition seemed a frail guide in this strange new territory. Yet a few rules of the new game seem to have been accepted by consensus. The first was that space generally and architecture in particular were implicated in the actions of invisible forces. Some of these forces were connected with new conceptions of physics, while others were tied to operational facts such as circulation or connectivity. Yet others could be sensational or bodily. All of these forces could now be related more explicitly to form, through adjoint dimensions. The second rule was that design must be quantified, measured, and dimensioned in ways beyond typical orthographic projection to accurately reflect contemporary forces and systems. New forces demanded new visual and graphic idioms. Design was searching for its own X-rays—its proper means to reveal the hidden but precise behavior in the bodies of buildings.

The standard-bearers for that search were, at first, few and idiosyncratic. Before 1915, the rare early architectural texts which dealt with the fourth dimension attained a cult status. Linda Dalrymple Henderson recounts that the American architect Claude Bragdon's (1866–1946) *Primer of Higher Space*, a quasi-mystical 1913 text which developed intricate tracery patterns from hypercube projections, was used by Johannes Itten (1888–1967) in his Bauhaus studios, and there is some indication that Wassily Kandinsky and Paul Klee were familiar with Bragdon's work as well.[45] "Adventure with me," Bragdon begins, "down a precipice of thought, sustained only by the rope of analogy, slender but strong."[46] Bragdon's *Primer* is liberally illustrated; perhaps a third of its pages are devoted to projective graphics of four-dimensional objects, very much in the spirit of Jouffret or Stringham. Among designers, Bragdon's investigations seeded an early curiosity in the visual structures appropriate to a four-dimensional architecture.

In 1915, German physicist Albert Einstein (1879–1955) first presented the field equations of his general theory of relativity to the Royal Prussian Academy of Science, recasting reality as a system of Riemann's multivariable functions.[47] Perhaps in part due to the broader cultural conversation swirling around relativity, a much wider range of architectural critics, theorists, and practitioners began to consider hyperdimensionality in architectural terms. Those architects who confronted questions of hyperdimensionality—and, just as significantly, its symbolic encodings—did so along a spectrum: from mystical analogy reminiscent of Bragdon, to conceptual catalyst for Theo van Doesburg, and finally to attempts at direct formal operation by the Russian avant-garde.

By the second decade of the twentieth century, architects were avidly debating the implications of hyperdimensional architecture and diffusing key texts of its popularization through their own contemporary periodicals, such as L'Esprit Nouveau (1920–1925) and De Stijl (1917–1931). These periodicals became organs not only of knowledge diffusion but of the cultural normalization of scientific ideas in architecture. Edited by Le Corbusier and Amédée Ozenfant,[48] the twenty-eight issues of L'Esprit Nouveau announced themselves as a "revue international d'esthétique," an intentionally ecumenical anthology of developments across the arts and sciences. The masthead, at various times, declared its purview to include art, science, film, literature, music, poetry, typography, and even sports. As a publication, it was not exclusively or even primarily concerned with architecture but rather with the total tendency of what it saw as an emerging modern age. A series of articles within L'Esprit Nouveau concerned topics related to science and the philosophy of science, many written by the engineer Paul Recht, and included biological, chemical, material, medical, and cosmological themes. The journal also published articles explicating Einstein's theory of relativity shortly after the translation of several critical papers into French.[49] These included Paul Le Becq's review of relativity, which emphasized the synthetic nature of Einstein's work in the application of four dimensions to physics.[50] Poincaré was also duly invoked as a precursor in Einstein's spatial sensibilities and particularly his notion of time.[51]

Van Doesburg's *De Stijl* fully embraced hyperdimensionality as both a calculational paradigm and a conceptual ideal. Like *L'Esprit Nouveau*, it aspired to cultural breadth, touching on arts ranging from architecture to painting and sculpture. But in contrast to *L'Esprit Nouveau*, which published primarily critical or interpretive perspectives when it covered scientific topics, *De Stijl* reproduced primary sources relevant to hyperdimensional thinking, such as excerpts of Poincaré's work.[52] In the several essays he himself wrote for *De Stijl*, van Doesburg stitched together the spatial and temporal dimensions of architecture; for example: "The new architecture takes account not only of space, but also of time as an accent of architecture. The unity of time and space gives the appearance of architecture a new and completely plastic aspect (four-dimensional temporal and spatial plastic aspects)."[53] Van Doesburg seemed to suggest that hyperdimensional mathematics can transcend the limits of past conceptions of space:

> In a future period in the development of modern architecture the plan will disappear. The composition of space projected in dimensions by a horizontal cut (the plan) will be replaced by an exact calculation of the construction. Euclidean mathematics will no longer be of use to us, but thanks to non-Euclidean calculations in four dimensions, construction will be simpler.[54]

The geometry of hyperdimensional form was foundational in van Doesburg's thought and in the wider context of De Stijl, particularly as a means to create an objective visual language. He argued that his aspiration for an objective and timeless nonstylistic style was achievable by the equally objective means of measurement and computation: "One must not hesitate then to surrender our personality. The universal transcends it. Spontaneity has never created a work of lasting cultural value. The approach to universal form is based on calculation of measure and number."[55]

Hyperdimensionality tempted designers like van Doesburg with the possibility that buildings could be parameterized as multivariable formulas, integrating and correlating their various constraints in a symbolic framework. If drawings appeared at all, they were secondary outputs of this primary calculative process. The manipulation of these variables could replace or at least qualify drawing as design's ur-language, inflected or accented stylistically but maintaining an irreducible logical core. This premonition found natural and radical resonance in constructivism. Perhaps most paradigmatic was an obscure but revolutionary 1928 article by the Russian architect Nikolai Krasil'nikov from *Sovremennaya Arkhitektura*, the leading constructivist journal of the late 1920s, that ventured a vision of the future van Doesburg yearned for: "To put it mathematically, the form of every body is a complex function of many variables."[56] These variables could be enumerated, and the architect must solve for the building as if it were an unknown in an algebraic equation:

> The form of each building must be solved for: (1) convenience of internal connections; (2) lighting of the internal accommodation; (3) ventilation of the accommodation and streamlining of the walls; (4) thermal insulation: ... Through plotting maxima and minima in relation to each of these factors, we should arrive at that building form which diverges least from them.[57]

What was truly novel was Krasil'nikov's suggestion that a calculus-based optimization process could conform architecture to hyperdimensional reality, and that such optimizations could yet be expressed in specific mathematical drawings:

> A scientific theory of the calculation of form is possible ... with the application of mathematical methods of analysis. ... Nothing has been done until now on the analysis of the actual form of the building. ... Our task is to extend this by more advanced paths, to give a scientific-objective assessment of all the possible variants. ... The intuitive-graphic method of designing ... can be replaced by the mathematical-graphic—a process in which intuition does not fall away, but occupies its proper place.[58]

In the vision of van Doesburg and constructivists like Krasil'nikov, not only did architecture exist in a matrix of dimensions far greater than three, but only the mastery of these higher dimensions brought imagination and method into proportioned balance.

In Krasil'nikov's work, we see the first rigorous consequences of reframing architecture as a symbolic encoding of multivariable mathematical functions. In his papers, plots of these functions are as common as more typical architectural drawings like plans or isometrics. Visual and symbolic representation reached an uneasy equilibrium in which graphic and numeric, visible and hyperdimensional converged.

Dimensioned Virtualizations

In discussing the history of surveying techniques, the historian of mathematics Katie Taylor introduces the useful notion of "vernacular geometry," in distinction to more rigorous and abstract theoretical geometry.[59] For Taylor, vernacular geometry is the application of mathematics to the material and visible world, including the quantitative process of measuring it. Architects have unselfconsciously used both vernacular geometry—like surveying—and more abstract theoretical mathematics—like hyperdimensional geometry—to expand the drawings and encodings used to represent the architectural imagination. In so doing, they gave vision an evolving role, from arbiter to interpreter of reality.

Digitization tied together the strands of triangulated drawing and multivariable functions as well as vernacular and theoretical geometry into a technologically enabled version of Krasil'nikov's calculated architecture. The computer became a device for the demarcation of virtualized mesh representations and their investiture with vast new complexes of dimensions. During the 1960s and 1970s, the researchers of Harvard's Laboratory for Computer Graphics and Spatial Analysis grafted together the cartographic and the hyperdimensional into a digital paradigm of spatial encoding. Founded in 1965 at Harvard's Graduate School of Design and known primarily as the birthplace of GIS (Geographic Information Systems, a term for digital tools for geospatial analysis), the lab's remit was much broader. An essential thread of research in *The Harvard Papers on Theoretical Geography*, the lab's technical reports to the Office of Naval

Research, was the reciprocal interdependence of triangulated networks and hyperdimensional projective geometry. The lab conjoined the two, and the mesh as physical terrain gave way to the mesh as datascape. Triangulation and hyperdimensionality were mutually orbiting terms of conversation in the early synthetic work of design computation.

Today, the triangulated mesh is a digital vernacular, a ubiquitous mode of encoding geometry for computational design tools, photogrammetric documentation, and computer-aided manufacture. As has been traced here, the mesh geometries of the early twenty-first century draw on deep roots in the vernacular history of surveying and drawing, and build on insights from the early twentieth-century encounter with theoretical hyperdimensionality. Indeed, it is the hyperdimensional turn of early modernism that opened the door to parameterization and endows the digital mesh with a dynamic and transformable quality. In this reading, the vernacular and the theoretical exist in continuous and constant interaction with the imagination of architecture itself.

The elastic fabric of mesh geometries and hyperdimensional parameterizations allow us to map our most radical imaginings from virtual vision to inhabited space, through the lens of mathematical measurement and drawing. Meshes propose dimensionality, and by extension reality, as a fact which can be encoded, and thus regulated and manipulated. Since their endless pliability allows meshes to encode virtually any spatial form, they are a medium of translation that renders architecture interoperable with every kind of physical and mathematical entity. The myriad deployments of the mesh—as surveying document, as photogrammetric reconstruction, as concretization of structure, as digital encoding of form—mark it as a decisive step in the mathematical virtualization of architecture as pure data.

5   A Virtuality Atlas: Stereoscopic Drawing and Geometric Dream Space

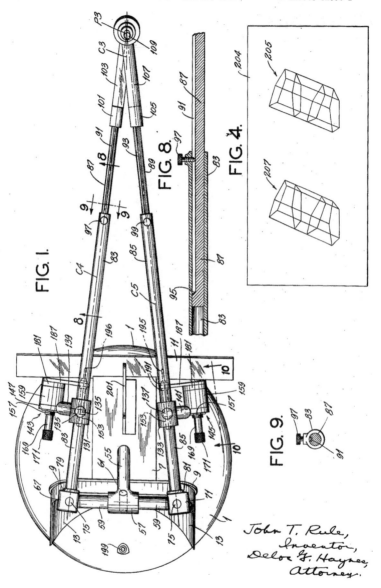

5.1   John Rule's unusual device for drawing wireframe stereograms.
Source: John Rule, "Apparatus for Producing Stereographic Drawings," US Patent 2,171,894,
filed November 17, 1937, and issued September 5, 1939.

fig. 5.1

If cartographic triangulation presented an objective and multifocal view of building and landscape at a politely detached remove, stereoscopy was an altogether more embodied affair: a technology to hijack human perception by hacking the rules of binocular sight. Stereoscopy—the simulation of an illusional three-dimensional view through two distinct perspectival images for each of the viewer's eyes—captivated both scientific and popular culture from the moment of its introduction in the mid-nineteenth century. The stereoscope was a keyhole into vividly simulated pseudorealities. As they peered through the carefully calibrated lenses of the stereoscope, users could behold two-dimensional drawings and images that ranged across seen and unseen dimensions. Stereoscopy announced a speculative medium in which imaginary mathematical spaces appeared immediately present and tangible.

Stereoscopy was arguably the deepest advancement in the technique of perspective since the architect Leon Battista Alberti's formulation of it in his 1435 *De pictura*. While art historians like Jonathan Crary have devoted considerable attention to nineteenth-century stereophotography, stereoscopic drawing and its particular expansion of the geometric imagination remain less understood.[1] In the century between 1850 and 1950, stereodrawing developed into a singular medium for scientists and visual experimentalists to behold speculative geometry. In distinction to the eidetic reproductions of stereophotography, stereodrawings were limited only by the ingenuity and imagination of the draftsman. Liberated from the burden of observed reality, stereodrawings could sketch entities and spaces that were invisible, theoretical, or even physically impossible. Moreover, the immediacy of stereovision appealed forcefully to the intuition of the viewer, beckoning her toward excursions in virtual territory. As it converged with the development of automated drawing machines in the twentieth century, it was stereoscopic drawing, not stereophotography, that was the clearest precedent for the first experiments in computer-generated virtuality.

The architectural implications of stereoscopy go well beyond the roots of virtual reality to the foundational understanding of mathematically drawn space. Stereoscopy was like a Copernican revolution in vision: by moving the projection plane from a detached distance to an attached proximity nearly coincident with the eye, geometry itself was implicitly moved from public objectivity to private experience. Yet for all its novelty, stereoscopy was not a schism from but a continuous extension of classical drawing techniques, as their inventor Charles Wheatstone argued: "It is worthy of remark, that the process by which we thus become acquainted with the real forms of solid objects, is precisely that which is employed in descriptive geometry, an important science we owe to the genius of [Gaspard] Monge."[2] The stereogram, the small rectangular cardboard slide that bore calibrated images for each of the observer's eyes, was not merely a recorder of binocular vision but also a sui generis format of information visualization. It offered a glimpse of a new and uncanny equivalence between geometrized data and space. When recognized at the junction of the real and the mathematical, scientific stereoscopy fused perception, geometry, and data in a new sensorium.

A Tourism of Virtual Forms

The origins of stereoscopy lay in a modest but curious 1838 paper by the polymathic English scientist Charles Wheatstone (1802–1875) entitled "Contributions to the Physiology of Vision: On Some Remarkable, and Hitherto Unobserved, Phenomena of Binocular Vision." Wheatstone observed that "the most vivid belief of the solidity of an object of three dimensions arises from two different perspective projections of it being simultaneously presented to the mind."[3] Surprisingly, he claimed, this simple fact seemed to have eluded the entire history of western painters and perspectivists. Based on this observation, Wheatstone developed the novel device he called a stereoscope: a tool for viewing two separate but coordinated images derived from two distinct projection planes, each one associated with one eye, displaced by a precise interocular distance of about 62 mm. The experience of the viewer using this new device was visceral and transporting: an illusion of depth and bodily presence was produced that was quite unlike a mere painting's monocular perspective, or any other kind of imagery which had come before. To accompany the stereoscope device, Wheatstone produced a series of drawing pairs which, when viewed binocularly, would fuse in the mind of the observer to reveal a scene or object in uncanny depth.[4] His stereoviews presented diverse subjects, including architectural motifs, and they became persuasive evidence for an entirely new way of seeing.

In their heyday in late nineteenth-century Europe and America, binocular stereoviews depicted the whole world of objects from historic monuments to singular biological specimens, and of spaces, from heroic battlefield scrums to quotidian domestic scenes. Stereoviews became an experiential currency that could be bought and traded. Views were rarely isolated one-offs but instead typically belonged to a thematic series or collection. In fact, the cultural practice of stereoscopy revolved as much around collecting as viewing, and these simulation collections constituted a new kind of experiential codex, a compendium of perceptual states. In 1859, remarking on the pervasive documentary power of stereoscopy, Oliver Wendell Holmes Sr. ventured:

> The consequence of [stereoscopy] will soon be such an enormous collection of forms that they will have to be classified and arranged in vast libraries, as books are now. The time will come when a man who wishes to see any object, natural or artificial, will go to the Imperial, National, or City Stereographic Library and call for its skin or form, as he would for a book at any common library. We do now distinctly propose the creation of a comprehensive and systematic stereographic library, where all men can find the special forms they particularly desire to see as artists, or as scholars, or as mechanics, or in any other capacity. Already a workman has been traveling about the country with stereographic views of furniture, showing his employer's patterns in this way, and taking orders for them. This is a mere hint of what is coming before long.[5]

In practice, this total library of views was replaced with a more modest sequential tour, and a simple case or shelf substituted for a vast library. Yet the result was the same: stereoscopy serially conquered the seen world and made the unseen world visible.

The wonderful playfulness at work in Wheatstone's stereoscopic drawings unintentionally encouraged a distinction between a perceived illusionism of stereodrawings and an alleged objectivity of stereophotographs. In 1852, the *Illustrated London News* reported "so long as mere drawings by hand were used, it might be held that the effect, however wonderful, was but some trick of art by which the senses were cheated. But the Daguerreotype admits no trick."[6] Furnished with photographic content, the stereoscopes of Wheatstone, David Brewster, and Oliver Wendell Holmes Sr. were among the most engrossing entertainments of the second half of the nineteenth century.[7] No fashionable parlor was without one. Since photography itself was also a relatively new technology, the stereophotograph was doubly novel, a revelation of verisimilitude and tangible presence. Early users of the stereoscope exulted in its ability to create the illusion of true-to-life *solidity*. They termed stereophotographs "solid daguerreotypes."[8] In 1859, Holmes observed: "A stereoscope is an instrument which makes surfaces look solid. All pictures in which perspective and light and shade are properly managed, have more or less of the effect of solidity; but by this instrument that effect is so heightened as to produce an appearance of reality which cheats the senses with its seeming truth."[9] By cheating the senses, stereophotographs gave documentary images a newly immersive quality, as well as amplifying the general fascination with the representational space they accessed. The photograph satisfied the deep fin-de-siècle craving for what Peter Galison called "a neutral and transparent operator that would serve both as an instrument of registration without intervention and as an ideal for the moral discipline of the scientists themselves."[10] The stereophotograph was a mute witness, noting with complete fidelity all of existence. It offered an earnest truth of dispassionate but transporting documentation.

For the European and American middle class of the nineteenth century, paired stereophotographs provided convincing vistas of remote locales, historic structures, and marvels of nature. The stereoscope became more than a parlor entertainment: it became an appliance for a particular kind of virtual tour. The historian Jill Steward notes the range of media that were available by the late nineteenth century to inform the literate public about where they should travel and what they should see: "While it is undoubtedly true that the emergence of new forms of visual culture, allied to popular entertainments in the form of dioramas and panoramas, played a major part in familiarizing the public with faraway places and making them attractive and fashionable, so too did the circulation of illustrated printed materials using the new reproductive technologies."[11] Stereoviews took full advantage of such technology, and toward the turn of the century, boxed sets for stereophotographic tours of specific cities and regions across Europe, Africa, Asia, and North America became widely available.[12] Stereoviews acted as both a televisual portal and an editing device, instantly presenting precisely the correct view of an exotic locale. In 1897 Underwood and Underwood, the American firm that became the largest stereogram producer in the world, led the charge with their boxed sets of Rome, London, Paris, and other destinations, distributed as the Underwood Travel System.[13] Like the instruments for mathematical surveying, the stereoscope could capture and document the world in an exchangeable medium. But unlike surveying, stereoscopy sought to present an embodied experience of space. The stereoscope in general, and these touring sets in particular, essentialized in stowable form the nineteenth-century impulse toward encyclopedic documentation not of facts or processes (as with Diderot's *Encyclopédie*) but of experiences, presented in the guise of an artificial (and pleasurable) tour. As Thomas Hankins and Robert Silverman have observed: "The stereoscope domesticated all Earth. By stereographically capturing the visual form of even the most exotic locations, photographers neatly analyzed and preserved the entire expanse of the planet for the benefit of civilization."[14]

fig. 5.2

The extraordinary new experience of stereoscopy reframed the contours of vision well beyond the recreations of virtual tourism. In domains such as science and medicine that prized meticulous observation, stereoscopy made inroads as a form of exact observational simulation. In fact, many saw surprising parallels between the experience of the tour and the researches of a pioneering scientist:

> The tourist cannot find a more faithful and true recorder than the stereoscopic camera, for its records have all the appearance of solidity and life. The scientific explorer

5.2   A slide from Underwood and Underwood's Stereograph Travel System, illustrating the
tour map used in tandem with the stereoscope.   Source: Underwood and Underwood, U-112648,
1908, stereoview, Boston Public Library, Print Department.

Works and Arlington, N.J.

SCULPTURE 1901 TRADE MARK

Studios Westwood, N.J.

61-Traveling by the Underwood Travel System—Stereographs, Guide-
oks, Patent Map System. Copyright 1908 by Underwood & Underwood. U-112649

ROME

cannot find a more faithful recording servant than the stereoscopic [camera]. No flat pictures here: all is seemingly life and solidity. The surgeon hands down to other generation[s] the life-like pictures of his operations. And so all along the line objects are represented as they should be, true to life.[15]

Parallels drawn between touristic impulses and scientific encyclopedism were not innocently rhetorical. Scientists built stereoscopic tours of their own objects of study, observational atlases of their specific scientific disciplines. Stereoscopic atlases ranged across clinical practice (1898),[16] anatomy (1906),[17] dermatology (1911), plastic surgery (1919), macular diseases (1928), and neurology (1947), to name only a few. Stereoscopy became a lingua franca of scientific collecting.

Lorraine Daston and Elizabeth Lunbeck have claimed that this mode of scientific observation "cultivates the senses of the connoisseur and straitens the judgement of the savant."[18] In this sense, the stereophotographs that were byproducts of scientific practice allowed the viewer to see both *what* and *how* the dissecting anatomist, the collecting naturalist, or the experimenting physicist would see. They were records not only of observed objects but of the act of observation itself. In establishing both common observations and common imaginations, they surmounted the gaps of understanding embedded in subjectivity.

fig. 5.3 A particular strain of scientific stereoscopy attended to objects at the frontiers of advanced geometry. In the stereoscope, complex figures of warped surfaces and crystal forms could suddenly be spaces of dwelling. Consider, for instance, the series of stroboscopic stereophotographs that Étienne-Jules Marey introduced in his 1894 book *Le mouvement*. Marey's stroboscopic flashes rhythmically captured instantaneous moments of objects in continuous motion, constructing a proto-cinematic time-lapse of images. While most famous for recording and analyzing the kinesthetics of humans and animals, Marey applied his chronophotograph technique of stroboscopic stills to mechanical movements as well. When captured by two synchronized cameras, the taut traces of thin filaments mechanically driven through air created the impression of hovering virtual volumes. Marey describes his process thus:

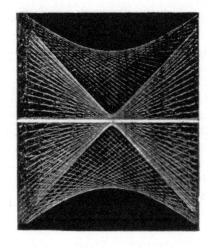

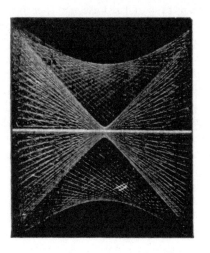

5.3   A hyperboloid of revolution and its asymptotic cone.
Source: Étienne-Jules Marey, *Le mouvement* (Paris: Librairie de l'Académie de Médecine, 1894), 28.

Let us suppose that the straight line, as it moves in space, leaves a record of its track at every point which it successively passes. Now, this purely imaginary supposition may become an accomplished fact, thanks to photography. Indeed, supposing we take a series of instantaneous views of an illuminated thread as it moves in front of a dark screen, figures are produced which exactly resemble the stereoscopic forms obtained by stretching a series of threads between metal armatures.[19]

Marey had recreated, stereoscopically, the intricate mathematical wire models of ruled surfaces developed by Monge's student Théodore Olivier. Marey observed: "By means of threads stretched between metal armatures, one can show how the successive positions of a straight line can produce cylinders, cones, conoids, and hyperboloids by revolution."[20] Stereoscopy held extraordinary capacities as a tool for the investigation of mathematical shape.

Remarkable as stereophotography was for recording geometry, its capacities as a speculative medium were limited. If stereodrawing could be dismissed as a trick of the eye, architectural researcher Richard Difford points out that the stereophotograph was criticized due to a "fear that it might become too truthful."[21] By comparison, stereodrawing proved a generous means to spatialize unseen or theorized forms. Its use thus inexorably expanded beyond documentation to encompass the representation of complex geometric ideas across the natural and exact sciences. Projective geometry, spherical trigonometry, and differential forms could all be convincingly, even viscerally, represented through stereodrawings. Beyond mathematical figures, stereodrawings conveyed diagrams of chemical, mineral, or even atmospheric and magnetic phenomena. Stereoscopy became a marvelous new conduit by which abstract concepts and even information itself were brought into the realm of the human eye.

Atmospheres Made Solid

While stereophotographs offered recorded factuality, stereodrawings fostered tangible speculation. In this capacity—as tools of speculative imagination—stereodrawings had their most subversive impact on visual culture. Unlike stereophotographs, stereoscopic drawings freed vision from the shackles of verisimilitude and accuracy and instead leapt into unseen, imagined, or distorted realities. If stereophotographs simulated the instantaneous raster experience of solid presence, stereodrawings attended to the spectral and uncanny experiences of vision that were only possible through a more graphic abstraction. Ambiguous optical illusions were one case in point: a fascination with warped optical effects that were only accessible through wireframe configurations was present in the earliest stereoscopic investigations. In fact, the kind of peculiar inversions of normal vision that were uniquely possible with linework drawings were essential to Wheatstone's conception of the stereoscope as an experimental instrument.

Wheatstone was intrigued by a famed 1832 problem of Louis Albert Necker (1786–1861), a Swiss crystallographer who first pondered the strange nonorientable quality of a wireframe cube. Necker recounted:

> The object I have now to call your attention to, is an observation which is also of an optical nature, which has often occurred to me while examining figures and engraved plates of crystalline forms: I mean a sudden and involuntary change in the apparent position of a crystal or solid represented in an engraved figure. . . . I have been a long time at a loss to understand the reason of the apparently accidental and involuntary change which I always witnessed in all sorts of forms in books on crystallography.[22]

Necker noticed a strange flipping of isometric wireframe drawings between concave and convex. Wheatstone was able to induce and maintain similar effects, which he termed "pseudoscopic," by switching the images intended for the left eye with those of the right, and vice versa, in a stereoview. This hack created the impression of a world turned inside-out, of space itself reversed. The sensation was akin to seeing the surface of a bas-relief from behind. Solid objects appeared hollow, and a kind of anti-solid total inversion dominated the visual field.

Besides being used to experiment with the optical oddities suggested by the Necker cube, stereoscopic drawing was also perfectly suited to represent the strange spatial forms of invisible forces. Stereodrawing rendered the spaces of ephemeral atmospheric, electromagnetic, and microscopic structures not only visible but immersive. It supplied the imagination with images that the technically unaided eye could not see. The speculative possibilities of stereoscopic representation were on vivid display in applications from geometric drawing, molecular chemistry, and planetary astronomy. Each of these fields cast the stereoview as a nexus between observed and imagined reality, communicating arcane systems in a viscerally palpable way.

fig. 5.4

Nineteenth-century physicists seized on the advantages of stereoscopic drawing to communicate and debate theories of increasingly abstract forces. Stereoscopic drawings were tools of judgment and persuasion in the formation of scientific consensus. James Clerk Maxwell (1831–1879), the Scottish physicist who first characterized the mathematics of electromagnetic fields, fully exploited this capacity. As early as 1868 Maxwell's inimitable hand-drawn stereoviews presented a series of complex field, curve, and surface constructions including "stereograms of the lines of curvature of the ellipsoid; and its surface of centres; of the wave surface of Fresnel; of confocal and spherical ellipses, of concyclic spherical ellipses, showing the form of Laplace's coefficient of the second order, of twisted cubics, of Gordian knots . . . and of four forms of the cyclide."[23] What is perhaps most curious about these stereograms as physical artifacts is that they leave no trace of their actual construction process; they are devoid of construction lines, evident compass or divider punctures, or other vestiges of calculation. The mysterious methods by which Maxwell drew these views perhaps supports his reputation for divine intuition. No less a mathematician than Felix Klein, who was highly attuned to visual representation, admired this approach: "What does distinguish Maxwell to a great degree is a strong intuition, rising at times to divination, which goes hand in hand with rich power of imagination. For the latter quality much evidence can be cited: his predilection for diagrams, . . . [and] stereoscopic figures."[24] Spatial intuition and its connection to stereoscopic drawing proved a durable association. The psychological and representational capacities of the stereoscope were very much entwined.

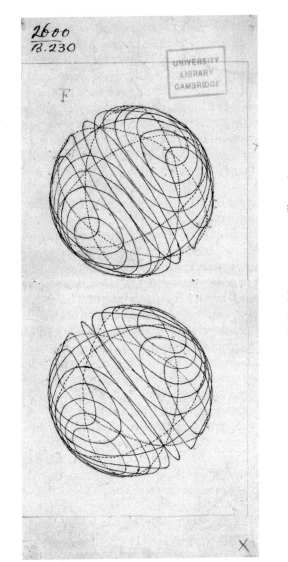

2600
B.230

UNIVERSITY
LIBRARY
CAMBRIDGE

5.4   One of James Clerk Maxwell's hand-drawn stereograms illustrating a structure of spherical curves.   Source: James Clerk Maxwell, "Stereographic Drawing," c. 1868, Stereoview, ms. Add.7655/v/i/11, James Clerk Maxwell Archives, Cambridge University Library.

fig. 5.5

Not to be outdone by physicists, chemists rushed to apply stereo techniques to the depiction of difficult-to-illustrate molecular and crystal reticulations. According to Klaus Hentschel, Paul Groth's 1876 *Physikalische Krystallographie*, an overview of crystal structures, was the first text to introduce such innovations to crystallography.[25] Yet perhaps the most influential early work did not arrive until some decades later, with Richard von Mises and Max von Laue's 1926 *Stereoskopbilder von Kristallgittern*.[26] Containing twelve painstakingly constructed stereograms documenting fundamental crystal structures such as cubic, hexagonal, and rhombic packings, it illustrated tangibly the expansive structures of typical crystal lattices. Other stereophotographic atlases of chemical and crystal models soon followed, such as the 1928 *Stereoscopic Photographs of Crystal Models*, edited by Sir William H. Bragg (1862–1942) and his son William L. Bragg (1890–1971), joint 1915 Nobel laureates in physics.[27] In this atlas, the Braggs recognized that stereograms hovered in a partial dimension between two and three. While they offered so much more depth than monoscopic two-dimensional drawings, stereograms inherited the single privileged view of perspective, and denied the viewer the infinite vantages of three-dimensional models. Yet in reviewing the Braggs' work, the crystallographer R. W. James saw that stereograms transcend the fragility and expense of cumbersome physical maquettes: "The ideal method, the study of three-dimensional models, is rarely available, but sets of stereoscopic models of this kind make a very good substitute for models, and one could wish that more of them existed."[28] Like Maxwell's stereograms, these views were really mathematical abstractions of force operations. They etched the hidden volumetric tessellations that organized the material world.

Beyond its unique ability to trace tiny or invisible structures, stereoscopy also became a surprising way to circumscribe dynamic systems which operated at vast scales. It was a way to imagine the shape of entire ecologies decades before they were truly visible through satellite remote sensing. Used in this way, stereoscopy removed the viewer from her terrestrial bounds and placed her at a vantage point that was only accessible through calculation. In his remarkably novel 1931 paper "Geophysical Stereograms," the German statistician Julius Bartels (1899–1964) proposed applying

Tafel XV.

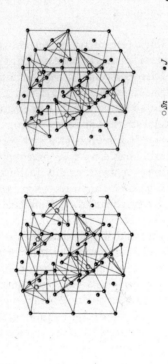

○ Sn   • J

SnJ₄, Zinntetrajodid.
Tin tetraiodide.

Verlag von Julius Springer in Berlin.

Gezeichnet von E. Rehbock-Verständig.

5.5   A stereogram of a centered cubic lattice, typical of stereograms representing
crystallographic structures.   Source: Richard von Mises and Max von Laue, "Thirring
Stereoskopbilder von Kristallgittern," *Monatsheft für Mathematik und Physik* 35, A38 (1928).
© 1928, Springer Nature.

stereoscopy at the scale of weather systems or planetary magnetic fields. Bartels expanded the concept of Maxwell's electromagnetic stereograms to the scale of Earth itself:

> Ordinary plane drawings and curves are inadequate for representing phenomena in three dimensions. This is keenly felt in several branches of geophysics, especially in terrestrial magnetism, where the changes in time and space affect all three components of the magnetic field-vector. . . . In order to meet this necessity, recourse was taken to stereoscopic drawings . . . which convey a vivid impression of the actual phenomena.[29]

Painstakingly constructed stereoviews laid bare the diaphanous formations of ephemeral and atmospheric phenomena. Bartels took matters a step past spatial representation, merging it with data visualization in a mutually registered virtual world. The stereoscopic drawing could encompass data constellations on a planetary scale, rendering abstract point clouds as encompassing spaces.

fig. 5.6   It must have been a startling awakening to understand that these cryptic images inscribed something fundamental about the comportment of physical reality. To interpret the intersection of a matrix of datapoints or a vortex of warped curves as real space was no trivial task; it demanded an inference of thick space from raw information. The geometry of invisible forces presented disorienting vistas which required trained spectators to fully consume. To properly train viewers in turn demanded a reassessment of technical education along stereoscopic lines, which inflected how knowledge from physics and mathematics to architecture and engineering was communicated in the early twentieth century. Most crucial for the design disciplines, the advent of stereoscopic projection reframed geometry as something to be experienced, not merely observed.

A Sensorial Training

As stereodrawing became a window into the imaginary spaces of theoretical science, educational reformers evangelized it as a door to radical new techniques for training scientific intuition.

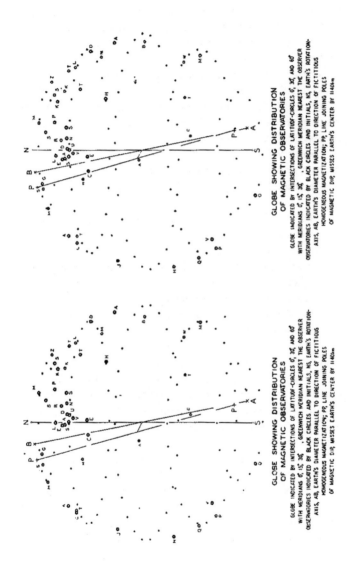

GLOBE SHOWING DISTRIBUTION
OF MAGNETIC OBSERVATORIES

GLOBE INDICATED BY INTERSECTIONS OF LATITUDE-CIRCLES 0°, 30°, AND 60°
WITH MERIDIANS 0°, 15°, 30° , GREENWICH MERIDIAN NEAREST THE OBSERVER
OBSERVATORIES INDICATED BY BLACK CIRCLES AND INITIALS, NS, EARTH'S ROTATION-
AXIS, AB, EARTH'S DIAMETER PARALLEL TO DIRECTION OF FICTITIOUS
HOMOGENEOUS MAGNETIZATION; PP, LINE JOINING POLES
OF MAGNETIC DIP, MISSES EARTH'S CENTER BY 1140km

GLOBE SHOWING DISTRIBUTION
OF MAGNETIC OBSERVATORIES

GLOBE INDICATED BY INTERSECTIONS OF LATITUDE-CIRCLES 0°, 30°, AND 60°
WITH MERIDIANS 0°, 15°, 30° , GREENWICH MERIDIAN NEAREST THE OBSERVER
OBSERVATORIES INDICATED BY BLACK CIRCLES AND INITIALS, NS, EARTH'S ROTATION-
AXIS, AB, EARTH'S DIAMETER PARALLEL TO DIRECTION OF FICTITIOUS
HOMOGENEOUS MAGNETIZATION; PP, LINE JOINING POLES
OF MAGNETIC DIP, MISSES EARTH'S CENTER BY 1140km

5.6   A geophysical stereogram that represents invisible magnetic fields spatially.
       Source: J. Bartels, "Geophysical Stereograms," *Terrestrial Magnetism and
       Atmospheric Electricity* 35 (1931): 187. © 1931 American Geophysical Union.

If stereoviews could distill Maxwell's most arcane theories to an immediately apprehensible virtual object, could they not also inculcate a refined geometric sensibility for a new generation? After all, using the stereoscope was as close as possible to seeing through the eyes of others to thereby arrive at common sensations. Stereoscopy held potential as a unique vehicle for the popular dissemination of complex spatial ideas far beyond rarefied scientific circles. It was a way to see what scientists—or architects—see: "From the works of the architect, the engineer, and the machinist, as exhibited in full relief, the student, whether at our schools or colleges, will derive the most valuable instruction."[30] Drawn stereograms became a tool for the discipline of visual intuition toward advanced geometric capacities.

Enthusiasm for stereo education reached a frenzy around the turn of the twentieth century, and Gilbert Pass, a teacher in Maidenhead, England, captured the general sentiment in 1900:

> I do not think one can over-estimate the value of placing a stereoscopic view before a class.... The pupil's interest is at once aroused, whilst some, too, are so beautiful that they would appeal even to the latent artistic taste of the average school-boy. The stereoscope may be used to correct erroneous impressions, and to convey perfectly accurate ideas.... In the hands of an enthusiastic teacher it can be put to almost limitless use.[31]

Advocates saw stereoscopy as a modern, visceral, and memorable mode of geometric training of students from children to professionals. Its impressions were so direct and vivid they appeared to implant ideas in the viewer with no instruction at all. A cottage industry of geometric stereograms soon bloomed to supply this new pedagogy.

A foundational application of stereo representation was to teach elementary solid geometry such as spheres, cubes, cones, and faceted volumes. From the earliest public discussions of the stereoscope, stereoviews of polyhedra and solid lattices had circulated as curiosities. By the turn of the century there was considerable optimism in mathematical circles that such aids could be turned to practical use in the wholesale transformation of geometric education. Several competing sets of stereoviews were developed to meet that demand. One of the most comprehensive was British mathematician Edward Langley's (1851–1933) *Solid Geometry through the Stereoscope* of 1907.[32] Langley's views exhaustively toured Platonic and Archimedean solids, as well as the various techniques for their drawing. Other mathematicians used anaglyphic presentations, with slightly displaced blue and red linework and associated glasses, to induce stereoscopic effect. French mathematician Henry Vuibert's 1917 *Anaglyphes géométriques* set the standard in this regard, presenting dozens of complex polyhedra in intricate anaglyphic drawings. Artists and designers found these representations irresistible. The architectural researcher Penelope Haralambidou has deftly shown that Marcel Duchamp found Vuibert's work mesmerizing and drew on it for his own subsequent experiments with stereoscopic drawings.[33] Both Langley and Vuibert, each in his own way, offered a grand tour of shape, aiming to circumnavigate the world of elementary solid geometry in a sequence of consumable views.

More advanced curricula treating geometric techniques like descriptive geometry and spherical trigonometry were developed to support professional training in engineering and architecture. Yet these more advanced techniques also required ever more exacting standards of draftsmanship. The drafting of two precisely coordinated views relied on rigorously skillful craft, and even small irregularities between the two views could shatter the brittle optics of stereo fusion. In the late 1930s and early 1940s John Rule, an MIT researcher, elaborated methods for the strict execution of intricate stereoscopic constructions such as descriptive geometry projections, bringing the connection to Monge full circle.[34] Rule describes the fastidious details such as scale, view orientation, treatment of curve intersections, and even text placement, which must be exactly realized in order to avoid shattering the illusion of stereoscopic solidity.[35]

Perhaps it was the unforgiving rigor of this method that persuaded Rule to build an elaborate and peculiar automatic device that encapsulated these intense processes in a single instrument, a radical new interactive machine for stereodrawing.[36]

fig. 5.7

figs. 5.8, 9

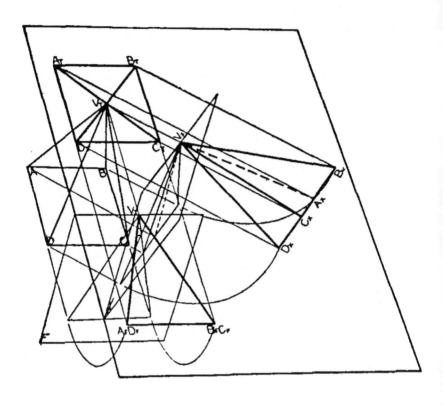

5.7 A stereoscopic descriptive geometry view, in this case the development of an auxiliary elevation.
Source: John T. Rule, "Stereoscopic Drawings," *Journal of the Optical Society of America* 28, no. 8
(1938): 321. © The Optical Society.

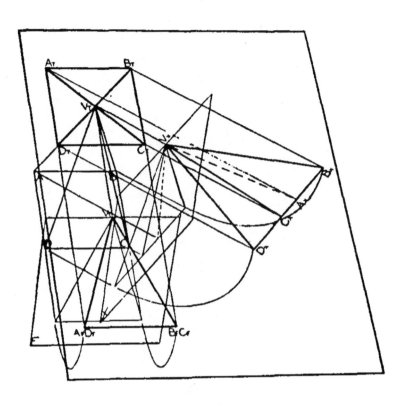

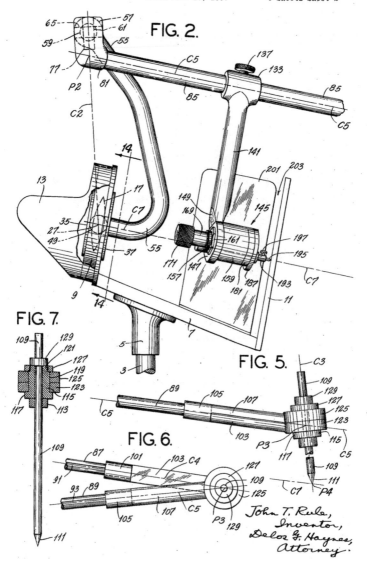

FIG. 2.

FIG. 7.

FIG. 5.

FIG. 6.

John T. Rule,
Inventor,
Delos G. Haynes,
Attorney.

5.8, 9　John Rule, "Apparatus for Producing Stereographic Drawings," US Patent 2,171,894,
filed November 17, 1937, and issued September 5, 1939.

FIG. 3.

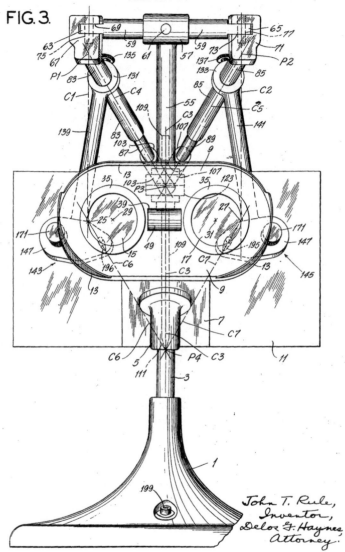

John T. Rule,
Inventor,
Delos G. Haynes,
Attorney.

Rule notes in his 1937 patent of the device that

> while stereographic photographs have long been used, the production of stereographic drawings has not heretofore been attempted to any great extent, primarily because of the complexity attendant thereupon.... However easy stereographic photographs have been to make, stereographic drawings, it will be seen, are fundamentally a considerably different proposition.... The apparatus of the present invention, it is believed, is the first apparatus that has ever been devised whereby stereographic drawings may be produced automatically.[37]

This improbable device equipped the user with three interconnected pens—one which was manually manipulated by the user's hand, and two auxiliary pens positioned automatically at the location of an imaginary projection plane at which stereopsis—the fusing of the left and right perceived images into a single three-dimensional image—would be attained. At this stereopsis plane was placed a sheet of paper on which the intended stereogram would be inscribed. The user perceived these two auxiliary pens as one, stereoscopically fused, synchronized in space, drawing an object as she moved the manual pen along a physical object. An engineering tour de force, it was a machine that seamlessly fused visual intuition and a technical means for intricate execution.

While utterly ingenious, the instrument's expense and complexity doomed it to remain purely conjectural. But as a device building on a lineage of sophisticated drafting and geometric tools, it constituted a high point for the reciprocity between subjective visual training and mechanized geometric procedure. In a single, cybernetic, human-machine symbiosis, spatial perception and generative drawing were combined. Rule's device was a radical new prosthetic to augment the geometric intuition of the subject. In the narrative of stereoscopy, it delegated part of the viewer's perceptive apparatus to an external device, inaugurating a machine that could systematically intervene in the relationship between geometry and vision.

Automated Stereodrawing

John Rule's stereodrawing device was one in a new library of instruments mechanizing the stereoscopic generation of form in the early twentieth century. While Rule's device was uniquely designed for human interaction, many others that developed in parallel were entirely automatic, drawing mathematical stereograms mechanically or electronically. They were binocular successors to the various curve-drawing machines of the nineteenth century, mechanically articulated to output stereoimagery of particular classes of forms. Due to the strict accuracy that stereodrawing demanded, automated methods to plot them with machined repeatability held irresistible attraction. Not only could mechanical or electrical machines execute coordinated drawings more rapidly than a trained draftsman, they could, in principle, be readily retuned to exhaustively permute virtually endless variations. Automatic stereoscopic machines could draw complete libraries of forms without human eye or hand.

Automated stereodrawing was particularly suited to plotting specific types of mathematical curves that otherwise had no tangible presence. It was a means to make the nonphysical tangible. Many of these machines drew on older research on the visualization of physical vibrations. The German physicist Felix Auerbach's (1856–1933) *Physik in graphischen Darstellungen* (Physics in graphical pictures; 1912) served as a précis text and prelude to the earliest stereoscopic automation. Auerbach was a visual popularizer of physical oddities, well known to architects and artists of the time. In fact, no less an architect than Walter Gropius completed Auerbach's house as one of the final projects of his Weimar office.[38] In the *Physik*, Auerbach introduced a suggestive method that used pairs of trigonometric curves displaced by a measured interocular distance for stereo viewing. Although Auerbach did not focus on their serial automation, his parameterization of trigonometric curves naturally lent itself to mechanical implementation. Through stereographic drawing machines, once obscure curves gained absorbing new legibility.

Charles Benham's marvelous 1928 twin harmonograph for drawing stereograms of trigonometric curves continued along Auerbach's path.[39] Benham, an English journalist and amateur inventor, experimented widely with the optical properties of kinetic bodies and particularly the drawing of figures from compound pendulum motion.[40] His device exploited the simultaneous drawing action of two connected but slightly displaced pens under the movement of carefully aligned pendula. The process was genetically related to drawing trigonometric curves, with the key distinction that Benham's curves decayed over time with the inevitable dampening of pendular motion. The two pendula perform a slow duet, their sweeping musical traces recorded in hyperbolically mannered pen strokes. Benham marveled at the phantasmal clouds that emerged from his device:

> A special interest attaches to them from the circumstance that, in spite of their realistic three-dimensional appearance, no such solid curves have ever had existence, though, if they are carefully traced and accurately mounted for the stereoscope, it is almost difficult to believe that one is not looking at a stereoscopic photograph of some intricate wire model of an actual solid curve.[41]

The stereoscope was Benham's window into a world of physically impossible entities.

Mesmerizing though they were, the slowly decaying motions of pendula were entirely unique and stubbornly inimitable. Their quality was akin to French poet André Breton's surrealist automatism, a reflexive trace of process as opposed to a controlled workshop of form. In contrast to Benham's more meandering and processional device, perfect reproducibility was the remit of more exact machines like Giambattista Suardi's eighteenth-century geometrical pen that functioned on the principles of geared permutation. These more replicable techniques not only drew endless and exact copies of specific virtual volumes but, through incremental tuning of their underlying mechanisms, could show morphing intertransformations in subtle and incremental gradients.

One device that explored the combinatorial capacity of stereoscopic drawing to striking effect was the "photoratiograph" of the British photographer Arthur Clive Banfield (1875–1965). Banfield was a fellow of the Royal Photographic Society and prolific inventor in micrography and cinematography. Around 1920 he developed his photoratiograph as an intricate clockwork for the simultaneous generation of stereo pairs. The mechanism resembles the parametrically variable rose engine, capable of drawing a vast range of distinct curves selectable through gear combinations. Using two synchronized beams of light, the photoratiograph etched pairs of intricate curves in tandem in dual photographic plates, displaced by a defined interocular distance to ensure their perceptibility as a single stereoscopic virtual volume. Here optical sleight-of-hand dances with geometric intricacy, drawing neon curves against a pitch-black background. The resulting wireframe tapestries have the haunting topological ambiguity of Klein bottles, Necker cubes, and other pseudoscopic curiosities.

figs. 5.10, 11

The controlled seriality of automated stereodrawings hinted at the astonishing possibility of stereoscopic animation. If frames could be generated in an incremental changing sequence, could they be stitched together as a continuous film? Such a sequence would be faithful to the logic of the tour, offering itineraries of continuously morphing figures. Yet, to fully realize such a prodigy, the animator must attend to both the science of projective geometry and the delicate formal choreography of ghostly virtual volumes. In sum, the task called for a hybrid character who was both engineer and designer to coax elegant form from the severe logic of stereoscopy. Roughly two decades after Benham's work, the Canadian-Scottish animator and graphic designer Norman McLaren (1914–1987) felicitously united those rare talents. He was an exceedingly methodical and technically astute animator, practicing what Michael Century has called "exact imagination."[42] McLaren effectively operated as a hybrid scientist-artist, his workspace functioning as both laboratory and studio. He created short films and sequences and published technical papers accompanying each detailing the process of its realization; Century comments on this process: "This parallel output is typical of successful practice-based research in general, where multiple outputs are produced along both artistic and technical-scientific lines."[43] In the charming short animations McLaren is best known for, such as *Dots* (1940), *Boogie Doodle* (1941), *Pen Point Percussion* (1951), and dozens of others, he synchronized technical and creative prototyping.

169

5.10 Arthur Clive Banfield's photoratiograph, 1920.
Source: Donation of Arthur Clive Banfield, Object Number: 1958-186. Science Museum /
Science and Society Picture Library.

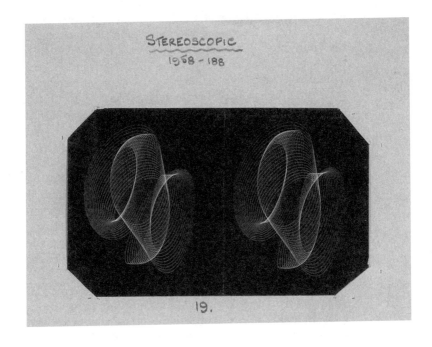

5.11   A stereogram (5 × 3 inches) produced by Banfield's photoratiograph.
Source: Donation of Arthur Clive Banfield, Object Number: 1958-188. Science Museum /
Science and Society Picture Library.

McLaren reveled in the ability of stereodrawing to concoct a new sensorium more attuned to graphic imagination. His overriding obsession was to use stereoscopy to construct realities uniquely expressible through the format of the drawing. McLaren hoped his films would "not simulate reality (a thing which natural stereophotography can do most ably) but create a new reality more in keeping with the graphic method by which the films were produced."[44] McLaren was carefully incremental in his work. Before fully attempting stereoscopic animation, he constructed several unusual stereograms and anaglyphs by hand. Film historian Ray Zone notes that McLaren was conversant with the technical literature on stereodrawing by pioneers like John Rule, and built on this technical history to develop his projects.[45] In the 1940s he drafted a series of contour-like stereodrawings, including a remarkable stereo portrait, and experimented with a stereopainting technique. One of his stereopaintings, *Honeycomb* (1946), depicts a universe dense with quasi-architectural tetrahedral volumes and hexagonal tessellations suspended above a blurred gradient horizon. McLaren's dreamlike drawings and paintings unfurled strange new worlds populated by wireframe architectures and luminous crystalline forms.

fig. 5.12

These excursions were merely a prelude to McLaren's far more ambitious project of fully stereoscopic animation. In 1951, during London's Festival of Britain, McLaren presented not one but two stereoscopic animations, *Now Is the Time* (1951) and *Around Is Around* (1951).[46] *Around Is Around* confronted directly the peculiar challenge of animated stereodrawing. McLaren wrote, "Stop motion of a solid scene shot by a stereo camera is indeed one solution to the problem of animated stereo film.... Our problem, however, was somewhat different, for we were concerned with the making of a stereoscopic film entirely from drawings."[47] *Around Is Around* permuted oscilloscopic forms of trigonometric curves through a sequence of carefully controlled waveforms. To achieve his sequences, McLaren adapted for animation precisely the methods used by Auerbach in *Physik* and by other experimentalists like Banfield. The film was a cultural adventure into a kaleidoscopic menagerie of forms inherited from scientific stereoscopy.

McLaren's animations were an apex of mechanical stereodrawing, supernal triumphs of his quest to depict the scopic

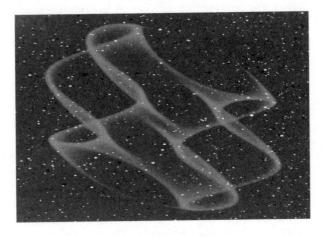

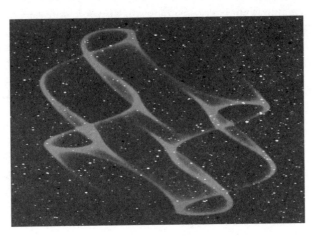

5.12 Still from Norman McLaren's film *Around Is Around* showing the oscilloscope-generated virtual forms as a stereoscopic pair. Source: Norman McLaren, *Around Is Around* (1951; Ottawa: National Film Board of Canada), film.

effects of endlessly fluid surfaces. They succeeded marvelously in making mathematical drawing something not only tangible but alive. The resonance with later digital formalisms is obvious. McLaren's animate, ephemeral surfaces presaged, in many ways, the real-time, dynamically computed forms of computational design. The vital medium of film was an explosive new way to inhabit geometry that moved, agitated, and folded around the observer. A sensorial cocoon was the logical evolution of stereoscopy's essential potencies.

Digital Stereospace

McLaren's astonishing animations transported spectators to virtual worlds possible only through stereoscopic drawing. Yet, as absorbing as these films were, the game of stereovision changed beyond all recognition with the invention of the electronic computer. On December 30, 1958, IBM released its 7090 mainframe, which would become a workhorse of government and private research groups, as well as a key engine for computational stereoscopy. It enabled vastly more detailed and data-driven stereoscopic forms than ever before. New software was the linchpin, and the act of coding software became indispensable to the whole practice of scientific seeing. Fields that had embraced stereoscopic visualizations, such as structural chemistry, plunged deeper into computational stereovision through bespoke software. In 1965 Carroll K. Johnson, a researcher at Oak Ridge National Laboratory, released ORTEP (Oak Ridge Thermal Ellipsoid Plot Program), a FORTRAN-based software for the construction of molecular stereograms from X-ray diffraction information.[48] The code, optimized for IBM 7090 machines, was openly distributed to other research chemists and intended for easy modification. Johnson reassured those specialists whose skills in mathematized drawing this innovation might have readily displaced: "The program does not in any sense replace the experience of the crystallographic draftsman; it is only a way of implementing certain of his ideas."[49] ORTEP transformed chemical imaging and became a unique means to communicate the densely layered filigree of molecular bonds. In effect, it was a new kind of computational lab to automatically plot spatial atlases of chemical form.

Even more astonishing than deliriously intricate chemical views were the fantastic tableaus of untamed surface geometries that revealed themselves computationally. These intricate surfaces held a magnetic attraction for visual experimentalists of every description, among them the NASA mathematician Alan Schoen (b. 1924). Schoen turned the power of digital stereoscopy toward dozens of stereoviews of computer-generated figures in his historic 1970 catalog of triply periodic minimal surfaces.[50] These space-filling surfaces, which echoed the mathematical models produced a century before, had natural architectural applications as screens, walls, and volumetric structures. Schoen's work was also closely affiliated with stereoscopic crystallography, both investigating periodic and cellular partitions of space. But Schoen used stereovision to yet greater effect: through his binocular pairs he built up a vast lexicon of lyrical space-filling labyrinths. Ever the cataloger, he identified twelve entirely new species of triply periodic minimal surfaces with distinct formal properties.[51] Just as novel were Schoen's algorithmic techniques for deriving these forms. Like McLaren's, his technical research and his perceptual investigations were unified in a single spatial pursuit. Schoen's surfaces and others like them proved of pivotal importance for computational designers, and they became an inspiration to sculptors and architects of the 1970s like Erwin Hauer and Peter Pearce. Through his work a dazzling fusion of geometric structure and immersive vision heralded a new type of algorithmic design.

The computer liberated geometry from its frozen state in drawing to the animate ecstasy of film. With the energetic dynamo of the computer, stereoscopic views came alive through rapidly and automatically calculated sequences of images. Ever more complex volumes and spaces became not only visible but experiential. It is as a culmination of these tendencies that we can best understand computer scientist Ivan Sutherland's (b. 1938) 1968 paper "A Head-Mounted Three Dimensional Display."[52] This evocative title suggests that once-detached perceptual auxiliaries were beginning to colonize the body itself, attaching and integrating themselves irrevocably. In the paper, Sutherland detailed a helmet display which presented the user with a wireframe dreamscape of phantoms miraculously superimposed on the physical environment.[53] This so-called

Sword of Damocles produced what Sutherland termed a "kinetic depth effect": "The image presented by the three-dimensional display which changed in exactly the way that the image of a real object would change for similar motions of the user's head."[54] There was complete synchronization between virtual space and kinesthetic space. Through an ingenious hack of stereovision and physical space, the constructed realm of descriptive geometry and the visceral reality of ocular vision met in a vertiginous digital orchestration.

Technically, Sutherland's headset display fused geometric calculation with classical optical techniques. It was composed of two small cathode ray tubes (CRTs) of one half-inch square each, driven by a custom-built matrix transform circuitry.[55] Like a space-age camera lucida, the artificial imagery from these CRTs was superimposed on the user's natural perspective: "half-silvered mirrors in the prisms through which the user looks allow him to see both the images from the cathode ray tubes and objects in the room simultaneously."[56] The experience also resembled that derived from then-contemporary stereoplotters, elaborate optical instruments used to draw contour maps from stereopairs of aerial photographs. Like these stereoplotters, Sutherland's device distilled dense information into ephemeral points and lines of light, floating in thin air.

It was an awkward and ungainly contraption. Yet the heavy instrumentation afforded the programmer considerable freedom as to the virtual entities the viewer could see. The user could pivot within a three-foot-radius circle about a base point, with an additional vertical range of three feet. The user could tilt her head vertically through an angular range of about 40 degrees.[57] She could employ one of two distinct positioning systems, one a mechanical arm which directly attached the headpiece and a second which used ultrasonic positioning. The visual motifs were calibrated to these motions: "Thus the analog line-drawing display, transistorized deflection amplifiers, miniature cathode ray tubes, and head-mounted optical system together provide the ability to present the user with any three-dimensional line drawing."[58] By describing the wireframe entities that the user encountered as "drawings," Sutherland provides stereodrawing with the perceptual armature to become fully immersive and spatial. These entities shared the same sensational quality of structured transparency, their full ghostly wireframe visible. They were, in a very real sense, direct descendants of Necker's cube.

fig. 5.13

McLaren's struggle for animated new virtualities grew ever closer to fruition in Sutherland's project, and nineteenth-century stereovision completed its migration into a real-time, digitally generated environment. In some accounts, Sutherland's work would constitute a culmination, and indeed its influence is hard to understate, being the forerunner of the myriad iterations of head-mounted virtual and augmented reality displays that continue to proliferate today. In this argument, though, it functions more like an epilogue to the experimental history of mathematical virtuality, already marked by stereoscopic experimentalists. Sutherland's work was an heir apparent to the hybrid data visualizations and virtual geometry of stereodrawing rather than the matter-of-fact mimesis of stereophotography. What the user saw made no pretense toward the solidity of stereophotography but instead was more like data made tangible: "Our objective in this project has been to surround the user with displayed three-dimensional *information.*"[59] While aligned to a physical space, the project's aspiration was to create an abstract datascape. This intertwining of space and data made the two inseparable, almost interchangeable. Data defined space, space became data, and the two symbiotically became a new type of architecture.

The Stereoscope and the Waking Dreams of Geometry

"The faith in geometry leads to a magic of spatial arrangements, in which life's difficulties will be solved by positioning things properly and making prescribed movements."[60] Robert Harbison's aphorism celebrates the capacities of mathematical armatures to situate and choreograph subjective experience. Stereoscopy was an application of subjective choreography in the extreme. It used those constructed situations to present a geometrized reality weightlessly untethered from the constraints of gravity and context. The single inert eye of the beholder was replaced with the dual roaming eyes of the creative agent. Stereoscopic drawing presented an uncanny way for the imagination to range

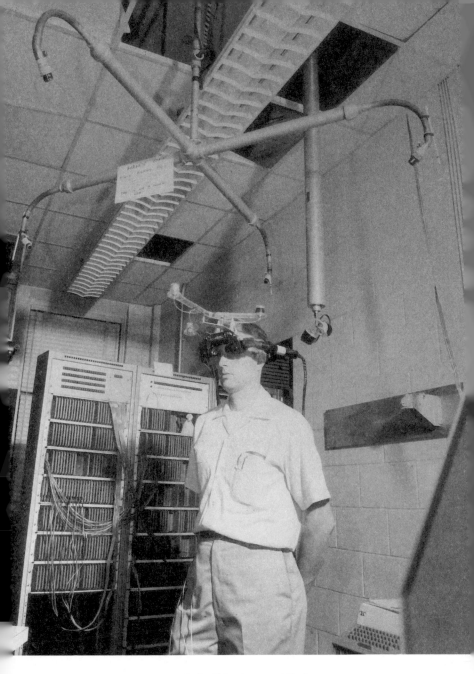

5.13   Sutherland's head-mounted display.
Source: Ivan Sutherland, "A Head-Mounted Three Dimensional Display," in *International Workshop on Managing Requirements Knowledge* (San Francisco, 1968), 760, 763.
Courtesy Ivan Sutherland.

across and through dimensions. Geometries which otherwise could be impossible to construct or even comprehend felt vivid and convincing through the lenses of the stereoscope.

Historians like Robert J. Silverman have justly emphasized the impact of photography, as opposed to drawing, on the cultural phenomenon of the stereoscope: "The advent of stereoscopic double photographs . . . extended dramatically the range of stereoscopic subjects, which had been limited to simple drawings like those contained in Wheatstone."[61] It is true that stereoscopic drawings burdened their authors with prohibitively laborious techniques that limited their production and diffusion. Yet that rarity and intensity only amplified their strange and wondrous characteristics. By placing no physical or conceptual limits on their objects, stereodrawings fueled a freedom of speculative representation that eidetic stereophotographs could not. The weightless and transparent wireframes of stereodrawings dissociated their presentations from the solidity of embodied reality, while still appealing to the more conceptual faculties of perception. When merged with automatic and particularly computational methods, stereoscopic drawing took on a new versatility and range impossible to match photographically. Stereodrawing was an agile auxiliary of creative imagination in a way that stereophotography never could be. Stereodrawings and the spaces they introduced hovered precariously on the boundary between true vision and pseudoscopic illusion. In a stereodrawing, it is often unclear whether what we behold is hyperreality or illusionistic fiction, an *illustration* or a *representation*. In the end, this borderland between fact and fiction is the native territory of the stereodrawing. The game of stereodrawing lies in the magician's sleight of hand between the two, and it is in this hyperillusionism that warped mathematics first became a habitable pseudoreality.

6   Illusion Engines:
Drawing at the Speed of Light

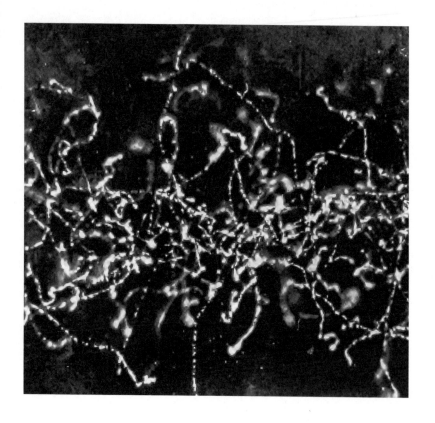

6.1   An early cloud chamber photograph.   Source: Charles Thomson Rees Wilson,
"On an Expansion Apparatus for Making Visible the Tracks of Ionising Particles in Gases and
Some Results Obtained by Its Use," *Proceedings of the Royal Society A* 87, no. 595 (1912), pl. 9.1.

In his very short 1911 paper "On a Method of Making Visible the fig. 6.1 Paths of Ionising Particles through a Gas," the Scottish physicist Charles Thomson Rees Wilson (1869–1959) unveiled two photographs that resembled faint pictures of a violent lightning storm.[1] Wilson produced these images in what he called a "cloud chamber," an ingenious invention for drawing the trails of subatomic particles and electromagnetic rays. Through a transparent aperture at the top of a cylindrical chamber with a radius of 7.5 cm and height of 5 mm, the observer could peer down on a strange atmosphere in microcosm. Inside, particles or electromagnetic rays ionized the supersaturated gas of the chamber, condensing minute silken clouds around their instantaneous paths through the cylinder. The vista below unfolded like a tempest in a bottle, crisscrossed by impossibly thin linear clouds. Historians Peter Galison and Alexi Assmus have noted that Wilson was "riveted by the phenomena of weather.... One must come to terms with the dust, air, fogs, clouds, rain, thunder, lightning, and optical effects that held the rapt attention of Wilson and his nineteenth century contemporaries" to situate instruments like the cloud chamber.[2] What was visible was not the subatomic particles themselves but rather the behavior—the weather—of their passage. With a tangential light in the otherwise black chamber, the intricate paths of particles and rays were illuminated in long-exposure photographic plates. These thin traces of kinetic traversal drew a portrait of a phantom realm with which the visible world was in intimate conversation.

In the first decades of the twentieth century, light drawings like Wilson's cloud chamber images—long-exposure photographs with figures drawn by flashes, pulses, and paths of light—were a graphic method that indexed motion across physics, engineering, art, and architecture. Specially constructed technical devices, in league with novel photographic techniques, plotted kinetic realities into complex composite images. This technique subverted the photograph's normally eidetic reproduction of light and replaced it with a diaphanous, cloudy, layered map of geometric motion. The foglike simultaneity of these images could seem illusional, and indeed the film historian Colin Williamson has compared them to sleights of hand or tricks of the eye.[3] But that simultaneity actually captured a precisely analytic representation of dynamic phenomena. Light drawing was a general tactic of temporal indexing that expanded how space itself was perceived.

Designers looked to tools like light drawing to visualize new virtual spaces and to bring speculative imagination into harmony with increasingly complex and inscrutable technologies. Hungarian designer and theorist László Moholy-Nagy (1895–1946) argued that "the multiplication of mechanical appliances, and new methods of research, required a new intellectual orientation, a fusion of clarity, conciseness and precision."[4] For designers like Moholy-Nagy, optical techniques such as the light drawing offered rigorous and legible graphic maps of physics itself.

Disciplinarity and Its Representations

By the early twentieth century, the meteoric rate of technoscientific advancement in Europe and America had far outstripped the capacity of any one person to comprehend it all. Moholy-Nagy lamented in 1928 that "the twentieth century overwhelmed man with its inventions, new materials, new ways of construction, new science."[5] Intellectuals of all persuasions struggled for a means to reconcile a vast and diverse technoscientific landscape with the yearning for a coherent humanistic experience. Physics in particular had leapt beyond the bounds of naïvely apprehensible realism into the far more remote territory of invisible behaviors. The invisible ranges of the electromagnetic spectrum were first artificially generated by German physicist Heinrich Hertz (1857–1894) in 1887, while the first subatomic particle was identified by the British physicist J. J. Thomson (1856–1940) in 1897. Other research, such as French physicist Henri Becquerel's (1852–1908) work on phosphorescence and radioactivity, questioned the boundary between light and matter. The physical world needed new representations, and light became a tactic to concretize unseen realities.

This explosion of scientific research coincided with the emergence of physics as a defined discipline and its intensifying specialization over the course of the nineteenth century. Before 1850, the term "physicist" was virtually unknown, and there were no regularly organized physics laboratories anywhere in the world. Historian of epistemology Peter Burke notes that the English term "scientist" itself was only coined in 1830, with "expert" and "expertise" also appearing in the nineteenth century.[6] The Clarendon Laboratory at Oxford University, the first real institutional laboratory, was established in 1868, and others soon followed at University College London and Cambridge University.[7] Thereafter academic physics departments became de rigueur in universities across Europe. Historian of science Josep Simon shows that in fin-de-siècle France and England, many aspects of these departments were regularized, including codified physics textbooks and instructional manuals that were distributed and used at an expanding scale.[8]

The development of physics as a discipline went hand in hand with the appearance of new graphic languages for recording scientific data. Physics of the nineteenth and early twentieth centuries was awash with lines, curves, traces, and graphic figures of every description, indexing a universe of physical phenomena. Graphic figures were the evidence in a relentless search for patterns in the material universe, and through indexed curves, these patterns were graphically sharpened into clear signals. German physicist Felix Auerbach's two closely related books, *Physik in graphischen Darstellungen* (Physics in graphical pictures, 1912)[9] and *Die graphische Darstellung* (Graphic representation, 1914),[10] were paradigmatic in this regard, each establishing representational conventions for a range of physical entities and actions including vibratory and oscillatory phenomena. Many such figures were plots of aleatoric, stochastic, or untheorized phenomena at the fuzzy frontiers of experimentation similar to Wilson's cloud chamber images. Historian of science Robert Brain observed that the "graphic method" of drawing and tracing physical phenomena was a "staple practice" for physics, particularly for mechanics but also for more theoretical fields.[11] The explosion of specialized graphic representation was part of the proliferation of specializations within physics itself. Amid the specialization of physics particularly and the sciences more broadly, intuitive graphic techniques that marked motion with lines and curves could, as Brain notes, create "a smooth disciplinary exchange of experimental results across institutional, international, and ultimately, disciplinary boundaries."[12] The circulation of graphic images eventually surpassed the boundaries of science itself to permeate the visuality of art, design, and architecture.

185

A Hunt for Universals

Design, like other fields, struggled with how to wrest a new unity from an explosion of knowledge, technique, and graphic representation in sciences like physics. Walter Gropius's memorable 1923 slogan "Art and Technology—a New Unity" provided an enduringly compelling, if necessarily reductive, aspiration to overcome this epistemic fragmentation through design. Through the Staatliches Bauhaus, the school he led beginning in 1919, Gropius aimed to realize this synthesis at the level of both method and consciousness. It was a rubric around which Gropius and his colleagues hoped to reconcile not only design with industrial manufacture but also the vast realm of scientific expertise with culture itself.

The "New Unity" of the Bauhaus was one of many early twentieth-century European projects to render scientific—and particularly physical and mathematical—knowledge more intuitively coherent. Each of these projects searched for some new linguistic, graphic, or sensory form of expression to stitch disparate fields together in common communication. The forerunner of this universalism was the fin-de-siècle attempt to construct a common language of science capable of expressing the truths of biology, chemistry, or physics in a mutually comprehensible way. The philosopher V. J. McGill (1897–1977) explained, "It is thought that the construction of a common language would enable scientists of various departments—physicists, biologists and psychologists—to exclude metaphysical problems and cooperate in the solution of the real problems, that is, those which permit of solution."[13]

For the cohort of physically minded philosophers known as logical positivists, symbolic logic offered the most rigorous basis for an intertheoretical epistemic language. With roots in the founding of the philosophical group Verein Ernst Mach in Vienna in 1928, the logical positivists advocated a universal language of science grounded in physics, mathematics, and ultimately the incontrovertible truths of axiomatic logic.[14] The original members of the group, such as the German physicist-philosopher Rudolf Carnap (1891–1970) and the Austrian philosopher Otto Neurath (1882–1945), drew inspiration from the foundational work of the German logician Gottlob Frege (1848–1925) and the subsequent extension of it through the *Principia Mathematica* (1910, 1911, 1913)[15] of British philosophers Bertrand Russell (1872–1970) and Alfred North Whitehead (1861–1947). In its aspiration to bring all of mathematics into a unified system, the *Principia* espoused exactly the kind of epistemic foundation and universal synthesis that the logical positivists craved. The *Principia* attempted to reduce the disparate expanse of mathematics, from non-Euclidean geometry to the theory of transfinite sets, to the rules, symbols, and syntax of deductive logic. For some philosophers, the *Principia*'s reduced and universal language of axioms provided beginnings of a "unified theory of knowledge" that went well beyond mathematics to span all of the sciences.[16] It could perhaps even be the germ of a "science of sciences."[17]

The logical positivist Otto Neurath saw a unified scientific language as a key to organize a total catalog of known science.[18] He posited that a vast ecumenical "encyclopedia will show that scientists, though working in different fields and different countries, may nevertheless cooperate as successfully within this wide field as when they normally cooperate within such special fields as physics or mathematics."[19] To realize that vision, in 1938, Neurath, along with Carnap and the American philosopher Charles Morris (1901–1979), inaugurated the monumental *International Encyclopedia of the Unity of Science*, a sprawling eight-volume series of monographs on the philosophy and sociology of science.[20] Though it was never completed, it aimed to be nothing less than a universally accessible catalog of all objective knowledge in a common scientific language.

Projects of epistemic unification like *Principia Mathematica* and the *International Encyclopedia of the Unity of Science* offered architects specific tools in the struggle for epistemic integration—namely, a deductive structure of first principles on one hand and a model of a common scientific language on the other. The search for unifying laws, languages, and grammars of design indeed became a durable preoccupation for architects, one which Gropius and other Bauhäuslers would persistently return to.[21] Yet the lexical tactics of philosophical positivism never quite matched the visual demands of design. Design entailed not only certain knowledge but also specific processes of perception, imagination, and creation. If there was a language to unify design and science, the graphic method of physics seemed closer to the way it might be written.

Instead of a panscientific symbolic language, designers like Gropius advocated visual and graphic systems as the vehicles of epistemic and perceptual unity. Moreover, for some at the Bauhaus, like Moholy-Nagy, technology—including the visual technology of photography and light drawing—could be co-opted as a tool for perceptual unification. The alignment of new machine media with new modes of visual perception presented entirely new opportunities for cultural and technical synthesis. Technology itself, with the graphic representations it engendered, could be a unifying language of design. Yet the overarching impulse of positivism—the unifying of knowledge in a common objective language—was legible in design's restless hunt for an objective grammar of vision.

A Precise Practice of Vision

Walter Gropius argued that the Bauhaus attempt to counter the atomic fragmentation of scientific specialization gone awry rested on its visual and cultural agenda. In his 1956 essay "Reorientation," he reflected:

> Our scientific age went to the extremes of specialization and has obviously prevented us from seeing our complicated life as an entity. This common dissolution of context has naturally resulted in a shrinking and fragmenting of life... . This disintegrating society needs participation in the arts as an essential counterpart to its atomistic effect on us. Made into an educational discipline—of which the Bauhaus was a beginning—it would lead to the unity of visible manifestations as the very basis of culture.[22]

Gropius hoped that a "unity of visible manifestations" across science, art, and culture could treat the malignant epistemic fragmentation of which ever more specialized expertise was a symptom. Such a unity could not be realized without an underlying unity of perception and its close corollary, measurement. In addition to expressing the need for a perceptual unification, Gropius and his colleagues, such as Moholy-Nagy, also believed that design's adaptation of quasi-scientific graphic methods had the potential to reintegrate creative consciousness. The aim was to counter the disintegration of narrow expertise with ways of seeing that were synthetic but technically advanced.

Gropius's phrase "the unity of visual manifestations" implied that graphic images—from the diagrams of architecture to the indexed curves of physics—could function as a common language between design and wider technoscientific culture. Yet this broad class of images included a wide spectrum of optical artifacts from precisely eidetic recordings to vague, ambiguous, or deliberately illusional constructions. These dual aspects of optical precision and visual subterfuge absorbed artists and designers of the early twentieth century. For Gropius, both the certitudes and pathologies of vision were essential to design technique. In an essay titled "Is There a Science of Design?" he argued: "I consider the psychological problems, in fact, as basic and primary, whereas the technical components of design are our intellectual auxiliaries to realize the intangible through the tangible."[23] As Gropius developed these ideas in more explicit terms, he contended that architecture must confront and manipulate the inescapable psychological parameters of vision. In March 1947 he attended a conference at Princeton University entitled "Planning Man's Physical Environment." In his lecture "In Search of a Common Denominator of Design Based on the Biology of the Human Being," Gropius proposed that "we have to study man's basic biology, his way of seeing, his perception of distance, in order to grasp the scale that will fit him."[24] There was even a hyperdimensional tint to this view:

> Our period has discovered the relativity of all human values. Accordingly, the element of time has been introduced as a new dimension in space, and it is penetrating human thought and creation.... Knowledge of the innumerable optical phenomena in space equips the architect to understand the interrelationships of voids and solids in space, of their direction, their tension or repose, and the psychological value of colors and textures. A solid foundation is thus laid on which many designers can rear a higher embodiment of creative unity.[25]

As Russell and Whitehead had tried to do in mathematics, Gropius was grasping for architecture's common ground with technoscience. He found it in laws of spatial and ocular perception.

Gropius emphasized that his approach to the Bauhaus idea included the development of a quasi-positivist syntax of vision. The underlying visual laws should be rooted in a quasi-scientific empiricism: "Intensive studies were therefore made at the Bauhaus to discover this grammar of design in order to furnish the student with an objective knowledge of optical facts—such as proportion, optical illusions, and colors."[26] The inclusion of optical illusions in this grammar of design reveals a surprising insight. Proportion and color had long been integral to design knowledge. As Moholy-Nagy affirmed, "colour should be valued as a primary biological law, just as necessary and indispensable for human beings as the fulfillment of other biological functions."[27] But to place optical illusions on the same level signaled Gropius's belief in the revelatory power of the pathological, the degenerate, the glitch. It suggested that any design epistemology should account for not only objective facts but also subjective perspectives. It declared a willingness to go beyond the limits of deductive or inductive logic to the operation of intuition and imagination.

The specter of illusion—such as the "illusion of motion," or the "illusion of floating space"—recurs in Gropius's writings. In fact, he maintained a collection of optical illusions in order to illustrate the primacy of perception in his lectures. His illusion collection consisted of accidents of proportional and color vision which could be orchestrated and manipulated by the designer. Gropius would draw on this collection in lectures to pose enigmatic design puzzles. His list of ten canonical illusions were all specifically architectural in character, sometimes dealing with details of construction, sometimes with overall composition. The illusions tended to be metarules of color and proportion. For instance: "Different colors differ in 'depth' and may cause, therefore, an apparent change in space proportions compared with the actual measurements of the space,"[28] or "A vertical line standing on a horizontal line of the same length appears to be longer."[29]

fig. 6.2

Beyond these ten examples, Gropius's writings offer other illusions which were more abstract but arguably more spatially experimental. Among the most telling was Gropius's illusion of a projected cube, a curious subversion of Albertian perspective. This illusion revealed that the perspectival wireframe, nominally a constructed and objective representation, could cloak strange

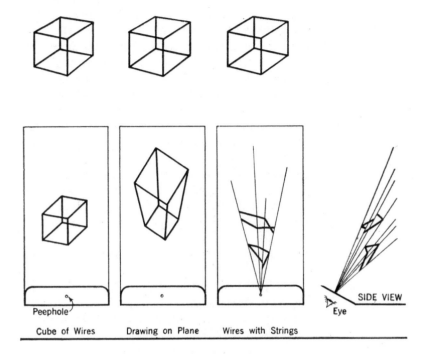

Peephole

Cube of Wires   Drawing on Plane   Wires with Strings

SIDE VIEW

Eye

6.2   One of Gropius's favored optical illusions, an instance of pathological depth perception. Source: Walter Gropius, *Scope of Total Architecture* (New York: Collier, 1962), 65. © 2020 Artists Rights Society (ARS), New York / VG Bild-Kunst, Bonn.

ambiguities. Gropius drew a bundle of eight wires, emanating from a point close to the eye of the observer, to which each of the eight vertices of a cube could be attached. The wires would trace the projections of the vertices of this cube from a specific projection plane, back to the origin point of the wires, and out to some arbitrary plane. Along these wires any of the vertices of the cube could slide toward or away from the eye, stretching and warping the true shape of the cube's edges while maintaining the same perceived shape from the perspective of the observer. A perspectival variant of the classic Necker cube, Gropius's clever illusion suggested wireframe transformations were an essential mode of illusionistic design research. It was an early intimation of the topology of impossible objects.

Gropius's perceptual research demanded both suitably trained designers and properly experimental instruments to interrogate and augment vision. The development of the requisite machinic intuition obliged the nurturing of a Bauhaus teaching faculty with a new hybrid set of skills, specifically versed in both design culture and the mechanical logic of technology. As Gropius recounted:

> The Bauhaus workshops were really laboratories for working out practical new designs for present-day articles and improving models for mass production. To create type-forms that would meet all technical, aesthetic, and commercial demands required a picked staff ... of wide general culture as thoroughly versed in the practical and mechanical sides of design as in its theoretical and formal laws.[30]

Gropius found that rare combination of cultural and mechanical inclinations in László Moholy-Nagy, who embraced a comprehensive synthesis of technology, vision, and psychology. At the invitation of Gropius, Moholy-Nagy taught at the school between 1923 and 1928, leading its *Vorkurs* preparatory course. His demeanor reflected a synthetic stance toward technology, as contemporaries recounted that Moholy-Nagy "chooses the workmanlike language of the technical expert: with words like 'working methods,' 'information,' 'specialist,' he deploys metaphors of the laboratory, R&D for the arts."[31]

fig. 6.3
Moholy-Nagy's research proposed kinetic forms of machinery as the elements of a new type of sensory experience appropriate to the modern age. His medium of choice for recording these

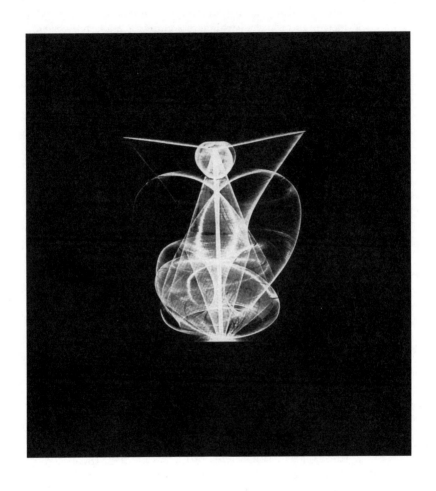

6.3   An example of a mechanically generated virtual volume that Moholy-Nagy considered the
highest form of sculpture, produced by his student R. Koppe in 1937.
Source: László Moholy-Nagy, *The New Vision* (New York: Wittenborn, Schultz, 1949), 47.
© 2020 Estate of Sibyl Moholy-Nagy / Artists Rights Society (ARS), New York.

kinetic forms was the long-exposure photograph, which unified innumerable discrete moments into a single fused vision of simultaneous time. These chronophotographs or motion studies were a kind of point-by-point plotting of action. Like triangular meshes encoding an irregular topography, motion studies were a perfect recording of an irregular gesture, a way of dimensionally collapsing spatial action into a single comprehensive portrait.

Though he was enamored of the formal qualities of motion studies, Moholy-Nagy's ultimate aim was to represent a dematerialized future of architecture and space. In his 1928 text *The New Vision*, he suggested an ordered sequence of stages for the evolution of sculpture that culminated, as a sort of transcendent apotheosis, with the modulated motion of motorized instruments. The sculptural space was not the machine itself but instead the traces of the machine's motion in space and time, the light drawing captured by a chronophotograph. Moholy-Nagy was fascinated by residual afterimages of rapid movement and the kinetic tracing of an implied virtual volume. For him, virtual volumes were a natural terminus of the modern impulse toward representation of space as an incorporeal body of pure light:

> In the lightening of masses, the next step beyond equipoise is kinetic equipoise, in which the volume relations are virtual ones, i.e. resulting mainly from movement of contours, rings, rods, and other objects.... With this transformation the original phenomenon of: sculpture = material + mass relationships, changes to the dematerialized and highly intellectualized formula: sculpture = volume relationships.[32]

The weightless forms of Moholy-Nagy's light drawings allude also to mathematical precedents like the wire models of ruled surfaces constructed by Gaspard Monge and Théodore Olivier. Though Moholy-Nagy was not an architect per se, he did speculate on the increasingly virtual volumes we might inhabit as culture and technoscience inexorably merged. His most tangible speculations were in his set designs for the 1936 film *Things to Come* (dir. William Cameron Menzies), which show a striking resonance of mathematical construct and perceptual representation. Comprising taut wire models and Plexiglas hyperboloids of revolution, the

fig. 6.4

194

6.4   László Moholy-Nagy, *Things to Come* models, 1936.
Source: Sibyl Moholy-Nagy, *Moholy-Nagy: Experiments in Totality* (Cambridge, MA: MIT Press, 1969), 128. © 2020 Estate of Sibyl Moholy-Nagy / Artists Rights Society (ARS), New York.

sets were the physical concretization of geometric light drawings. Regarding the sets for *Things to Come*, Sybil Moholy-Nagy recounted: "The fantastic technology of the Utopian city of the future would, so Moholy dreamed, eliminate solid form. Houses were no longer obstacles to, but receptacles of, man's natural life force—light. There were no walls, but skeletons of steel, screened with glass and plastic sheets."[33] The motion of light produced an illusion of solidity that was an idealized future state of architecture. It was the ultimate primitivism, a future architecture dissolved in radiant splendor.

Moholy-Nagy built a series of kinetic devices to produce symphonies of phantasmal light—moving beyond the chronophotography of virtual volumes to the dynamic and immediate motion of projected light. For instance, his well-known *Light-Space Modulator* (sometimes called *Light Prop*) of 1930 was a complex kinetic light projector which produced luminous effects as animated spatial sequences. Composed of a thin metal frame connected to a motorized mobile of mirrored and perforated metal sheets, plastic filters, and assorted spindles and rods, the modulator created tumbling, kaleidoscopic light patterns cast onto the walls, ceilings, and floors of the host room. Intended for ballet and theatrical productions, it evoked the effect of entering a film projector, or seeing what the projector might see. As it danced and gesticulated, the shimmering shards of light that enveloped the room dematerialized the space and gave spectators a glimpse of an infinite cosmos of illuminated form. The prop's physicality evaporated into borders and boundaries of pure luminance. Indeed, Moholy-Nagy sometimes referred to the light prop as his "space kaleidoscope."[34] Rosalind Krauss noted the prop used "the radiance of electric light to undermine the physicality of the object which is the source of that radiance."[35] Moholy-Nagy pursued an intellectualized notion of space that paradoxically relied on physical intermediaries such as his light props and set models. In this sense his practice correlated with the mathematical use of physical wire geometric models as proxies for more abstract spatial constructs of mathematical entities themselves.

Moholy-Nagy's language of machinic light drawing converged with wider projects of intellectual and scientific unification with his founding of the New Bauhaus in Chicago in 1937. The

fig. 6.5

school embraced a quasi-scientific experimental mindset that echoed Gropius: "This is not a school but a laboratory in which not the fact but the process leading to the fact is considered important."[36] Like the Bauhaus before it, the New Bauhaus aimed for a unity of experience and knowledge through perception. Though it operated under that name for less than a year, Moholy-Nagy quickly reestablished it as the School of Design (later renamed the Institute for Design).[37] Each incarnation of the school aspired to an intellectual unity between design, science, and technology; and the school's professor of intellectual integration Charles Morris, a colleague of Otto Neurath and a positivist philosopher himself, was the clearest spokesperson for that conceptual ambition. An indefatigable advocate of the "unity of science" movement, Morris had deep affiliations with Neurath's *International Encyclopedia of Unified Science*, which reverberated through his approach to the New Bauhaus and its successors.

In his 1939 paper "Science, Art, and Technology," Morris articulated the unifying spirit of the New Bauhaus:

> A consideration of science, art, and technology is inevitably a study of basic human activities and their interrelations. The theme is an old one, and many pages by many writers in many centuries supply variations and commentary. Yet the theme is a timely one for an America that stands before decades of high promise and peril, and for this Review which has recently opened its pages to a discussion of the place of the arts in the unity of science movement.[38]

For Morris, the issue of a common and universal language of art and design was a natural extension of the positivist project of a common scientific language. In fact, the language of science could be extended to encompass art itself. Morris reiterated in the first catalog of the New Bauhaus that "science, and philosophy oriented around science, have much to contribute to realistically conceived art education in the contemporary world.... We need desperately a simplified and purified language in which to talk about art in the same simple and direct way in which we talk about scientific terms. For the purpose of intellectual understanding art must be talked about in the language of scientific philosophy and not in

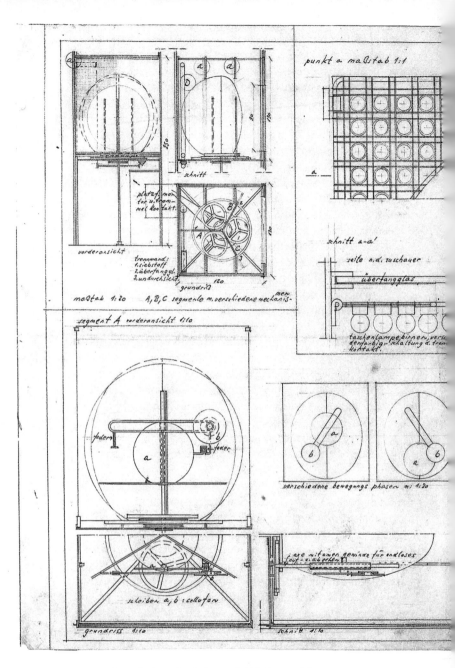

6.5   Schematic plans of Moholy-Nagy's *Light-Space Modulator*, 1930.
Source: Bauhaus Archiv, Berlin, Germany 2410; Harvard Fine Arts Library, Digital Images &
Slides Collection d2017.01187. © 2020 Estate of László Moholy-Nagy /
Artists Rights Society (ARS), New York.

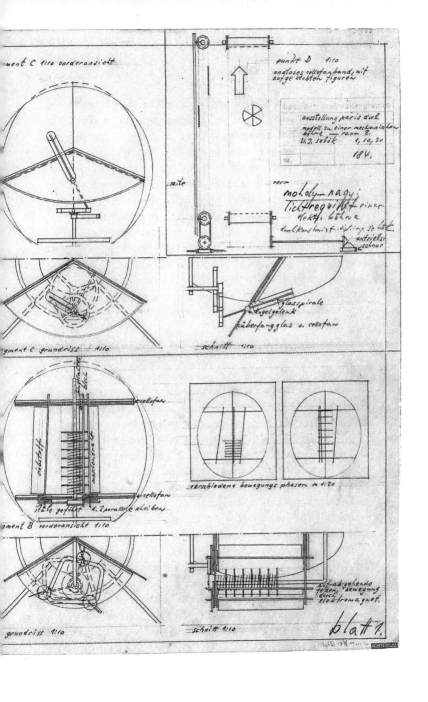

the language of art."[39] This hoped-for language—a grammar of art and design—was a direct extension of both Gropius's search for the language of vision and the logical positivists' quest for a unified language of science.

Whether Morris realized it or not, Moholy-Nagy's techniques of machined light and affiliated graphic methods had already furnished an expansive visual language for unifying art, design, and sciences grounded in physics. Though Moholy-Nagy had directed them toward aesthetic ends, light-drawing techniques were genetically related to ways of visualizing physics that were already well established by 1939. Whether for the evidentiary dimensioning of motion or for imaginatively etching portraits of strange weightless spaces, a generation of experimentalists embraced the graphic automatism of light projection as an exact language of vision. As if they were carefully tuning delicate and distant signals, these experimentalists modulated continuous, variational, parametric dimensions to conjure phantasmal virtual volumes at the boundary of physics and design.

Light Drawing, Analytic and Generative

The light drawings that so captivated Moholy-Nagy offered a unique device to unify perception and imagination across disciplines and spaces. Light drawing became a path to exceed normally limited human faculties and provide a richer, denser, truer maps of physical realities in an immediately apprehensible graphic language. As a technique, light drawing oscillated between two poles: a first that adapted it as an analytic scientific instrument to measure and graphically index complex human, mechanical, and physical motions, and a second that reveled in its generative and aesthetic possibilities as a combinatorial engine for strange new spatial volumes. On the one hand, light drawings were tools to optically record the behavior of kinetic and physical systems as exactly as possible. On the other hand, they were part of a long tradition of machines that drew strange algorithmic virtual volumes as diaphanous spatial apparitions, more radiant vision than tangible architecture. Light-drawing and projection machines were thus devices for both dispassionately encoding existing spaces and imagining spectral new ones. As Colin Williamson observed of the similar practice of chronophotography, these machines sat on a spectrum "between modern science and modern magic."[40]

To excavate the twin lineages of analytic and generative light drawing, one must return to their mid-nineteenth-century technical origins in the automatic drawing of complex physical phenomena. Their common ancestors were the peculiar mid-nineteenth-century graphs of the vibrations of sonic resonance known as Lissajous curves. First documented by the French physicist Jules Antoine Lissajous (1822–1880) in 1857, these parametric equations or their close cousins appeared constantly in the visualization of rhythmically periodic vibrations.[41] The product of experiments into making acoustic phenomena visible, these curves were first drawn by Lissajous as paths of light reflected off of a vibrating tuning fork onto a photographic plate. They were, in effect, among the first of what Moholy-Nagy would later describe as virtual volume traces of mechanical kinetic motion. Their remarkable ornamental quality inspired experimentalists to devise ingenious new types of machines relying on pendula, compound gears, or vibrating armatures to produce endless permutations of light curves.

fig. 6.6

The oscillations and vibrations that Lissajous meticulously studied were intricate physical phenomena, difficult to calculate but perfectly suited to representation by graphic methods like light drawing. Later experimentalists like French scientific photographer Étienne-Jules Marey (1830–1904) embraced oscillatory phenomena as a paradigmatic case for graphic methods in the sciences.[42] Marey was best known for his prolific physiological chronophotography that traced the motions of human or animal movements like walking and flying. He also turned his attention to physical phenomena, proposing applications of the graphic method in ballistics, aerodynamics, hydrodynamics, wave propagation, and even quasi-architectural applications like the resonating frequency of bridges.[43] Marey's 700-page *La méthode graphique dans les sciences expérimentales* served as a catalog of graphic transcription across fin-de-siècle sciences, as well as an argument for the universality of the graphic method itself. For Marey, the graphic method could act as a kind of universal language:

> Science has before it two obstacles that hinder its progress: first, the insufficiency of our senses to discover the truth,

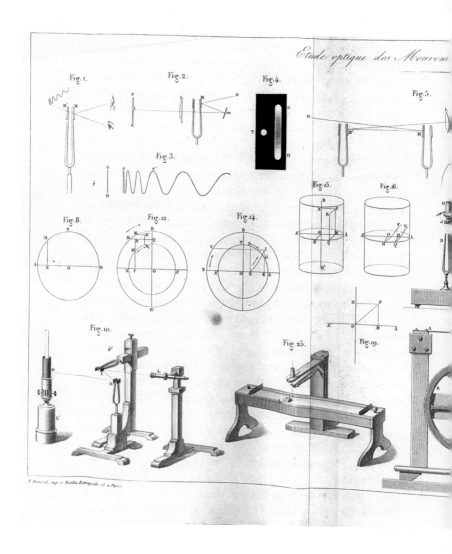

6.6 Lissajous's various devices for tracing modulated vibrations, as well as the trigonometric geometry for generating them.   Source: Jules Antoine Lissajous, *Mémoire sur l'étude optique des mouvements vibratoires* (Paris: Mallet-Bachelier, 1857).

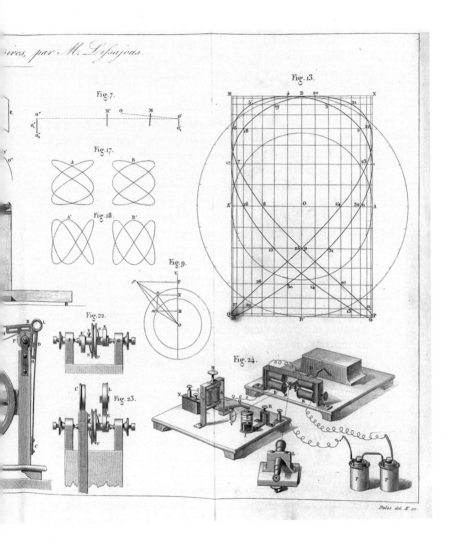

Fig. 7.

Fig. 13.

Fig. 17.

Fig. 18.

Fig. 9.

Fig. 22.

Fig. 23.

Fig. 24.

Dulos del. & sc.

and then the insufficiency of language for explicating and transmitting the truths we have acquired. The object of scientific methods is to dispense with these obstacles. The graphic method achieves, better than any other, this double goal. In effect, in delicate research, it captures the nuances that escape other means of observation; it acts to expose the processes of a phenomena, and then translates those processes with a clarity that language does not possess.[44]

Among Marey's light drawings were stereoscopic photographs of all manner of kinetic movements, from the human body in motion to the oscillatory traces of Lissajous curves. These images drew kinesthetic motion as wireframes of pure light.

fig. 6.7 The analytic application of light drawing to the physics of human motion was popularized in America by the business consultants Frank and Lillian Gilbreth in the 1920s. Taking inspiration from Marey's earlier motion studies, they recorded the virtual volumes carved from human action, using time-lapse stereophotographs of incandescent lights attached to the wrists, elbows, and heads of their subjects. Drawing on his background as a building contractor, Frank Gilbreth famously elaborated motion studies to deduce the optimal movements of a laborer in laying brick. He also went much further, applying the same technique to the motions of machinists, typists, and surgeons.[45] By varying the pulse and brightness of the tracing light, the Gilbreths could register a uniform segmentation of time and an unambiguous trajectory of motion of each of these actions. The resulting stereoscopic "cyclograph"—or light drawing—could be reconstituted as a physical "motion model"—"a wire model that exactly represents the path, speeds, and direction of the motion studied."[46] Since it collapsed the dimension of time into a single image, the Gilbreths sometimes referred to their method as "A Fourth Dimension for Measuring Skill."[47] The Gilbreths' goal was nothing short of the total optimization of kinetic work through "a means to permanent and practical waste elimination."[48] In effect, they considered the human body as a machine. Yet, beyond these pragmatic ends, the Gilbreths also suspected their light-drawing method was a transcendent new way of seeing that could unify the understanding of kinetic experience in a profound way. They

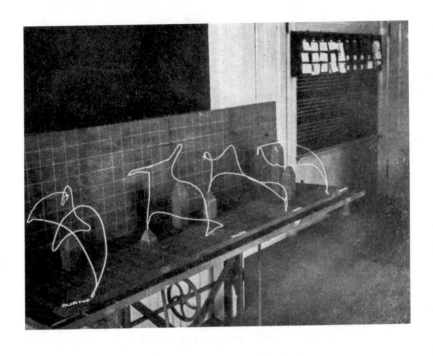

6.7 Frank and Lillian Gilbreth's wire models of "one man's progress learning paths of least waste" in the use of a drill press, transcribed from human action.   Source: Frank Bunker Gilbreth and Lillian Moller Gilbreth, *Applied Motion Study: A Collection of Papers on the Efficient Method to Industrial Preparedness* (New York: Macmillan, 1919), figure 16.

observed a "new feeling that is growing up, in all fields of activity, of the necessity of correlation," a reaction to "the tendency of this age to think in parts rather than wholes, in elements rather than grouped elements."[49] The motion diagram unified the dimensions of space and time through vision.

fig. 6.8

The architectural relevance of this technique was obvious. Motion studies demarcated the thresholds between human agents and the space that they occupied, carving an exact envelope of bodily motion. The Gilbreths' students Jane Callaghan and Catherine Palmer played through the architectural implications of motion traces to "measure space requirements of family functions in order to redesign the dwelling."[50] In their 1944 book *Measuring Space and Motion* they recorded the strange volumes drawn from the movement of human bodies during mundane domestic activities: putting on a shirt, or getting into bed, or brushing one's teeth.[51] These volumes, they argued, are the truest constraints of human habitation, the logically essential limits of architecture itself. Often set against the backdrop of gridded wallpaper, Callaghan and Palmer's studies were reminiscent of certain mathematical models of ruled-surface motion. They were volumetric snapshots of quotidian life. More fundamentally, they were skeletal figures of how space was occupied, graphed functions of how humans and architecture interacted.

fig. 6.9

Motion studies usefully registered the spatial behavior not only of sentient humans but also of autonomous machines. In 1952, while at Bell Labs, the electrical engineer Claude Shannon (1916–2001) developed Theseus, an electromechanical mouse which could "learn" the solution to an arbitrary maze puzzle through trial and error. The historian Ronald Kline noted that the apparatus was composed of around "seventy-five electromechanical relays of the type his employer, the Bell System, used to switch telephone systems."[52] Shannon thus fashioned a robotic nervous system with the components of the national nervous system of the telephone. To document the actions of this learning machine, Shannon attached a small light to the mechanical mouse and recorded motion studies of its iterative solution of the maze. This permutable maze was a mini-architecture for automatons. Partway between the conscious movements of the Gilbreths' subjects and the entirely determined motion of Moholy-Nagy's machines, the adaptive traces of Shannon's Theseus suggest the emergent and

6.8   Motion volume of a single man putting on a shirt.
Source: Jane Callaghan and Catherine Palmer, *Measuring Space and Motion*
(New York: John B. Pierce Foundation, 1944), 40, 44. © John B. Pierce Foundation.

6.9   Motion studies of Claude Shannon's Theseus mouse-maze machine.
Source: Path Through Shannon's Maze, 1952 (97-1526).
Courtesy of AT&T Archives and History Center.

intricate qualities of cybernetic feedback. They drew the artificial spatial perception and learning of the automaton in real time.

The scientific practice of motion study reached its zenith in the mid-twentieth century as it became increasingly central to the design of highly engineered environments and spaces. As machines grew to the scale of architecture, light drawing became a way to mutually mark and negotiate the activity of human and machine in this new cyborg symbiosis. Nowhere was that symbiotic interaction more critical than in the nascent field of spaceflight, and the associated realm of space habitats. From the beginnings of the Mercury manned spaceflight program and throughout the 1960s, NASA developed extensive motion studies to measure the dynamic range of its various spacesuits. What emerged was a sort of Vitruvian Man of spaceflight, diagramming the ergonomic range of a spacesuit's ideal motion.

fig. 6.10

NASA and its subcontractors applied motion studies both to the bodies of the astronauts and to the capsules and habitats they would eventually occupy. Light drawing became a way to document simulations that were both spatial and temporal. NASA's MASTIF (Multiple Axis Space Test Inertia Facility) apparatus was a tool for motion studies of a human-machine symbiotic environment, or in Moholy-Nagy's terms, a building-scale light prop. MASTIF was a huge three-axis gimballed motion cage used to simulate the flight of a tumbling spacecraft under partial control of an astronaut.[53] As it pitched, rolled, and yawed, MASTIF traced a legible path of the astronaut, indexing both the skill of the pilot and the simulated path of the capsule through space. Motion studies were also used in the planning of space habitats. As if revisiting the work of Callaghan and Palmer in an orbital context, NASA's contractors fashioned mock-up space station interiors that could be inhabited and traced by motion studies—the full-scale human version of Shannon's mouse maze.[54] The inscribed simulated motion of occupants could mold the geometry of the space environment itself.

fig. 6.11

Motion studies drew virtual volumes of human and machine life, from the quotidian context of the house to the extreme environment of the space station. These analytic light drawings were not speculations of a luminous future but portraits and quantifications of an invisible present. They were devices to level and make

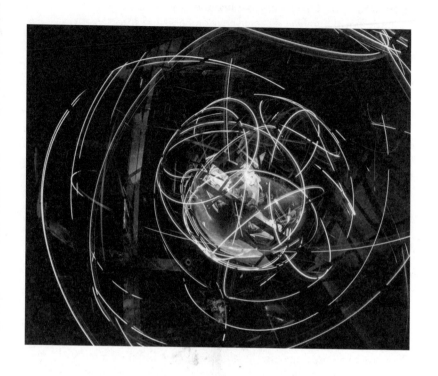

6.10 MASTIF, or Multiple Axis Space Test Inertia Facility, a NASA rig for motion-testing
Mercury astronauts.   Source: NASA, "Gimbal Rig in Motion," 1959, photograph, NASA Archives.

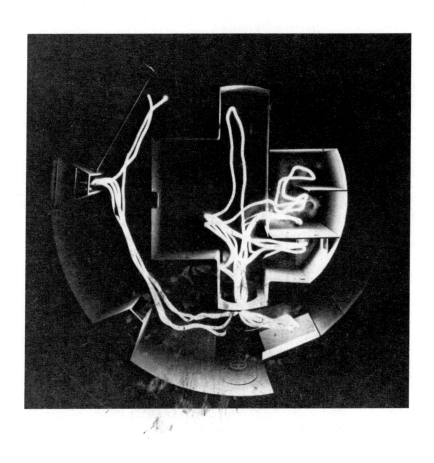

6.11   Motion studies within a proposed space habitat.
Source: Clovis Heimsath, *Behavioral Architecture: Toward an Accountable Design Process*
(New York: McGraw-Hill, 1977). © 1977 McGraw-Hill.

comparable all human, animal, and machinic behavior in a common graphic notation of time. In that sense, they achieved in very concrete terms the perceptual unification that Gropius saw on the horizon.

Aesthetic Engines

Analytic varieties of light drawing and the affiliated graphic methods of physics had an essentially referential function: they were secondary evidence of invisible entities or actions that were the primary object of interest. But the relationship of image and object could also be decoupled and inverted. While an image typically documents the behavior of an object, the image itself could also become the object of interest. Rather than being seen as a space of forensic evidence, the light drawing could be investigated as a space of creative generation. Light drawings, decoupled from their nominal objects of study, could trace spaces of speculative imagination.

Like its analytic counterpart, generative light drawing originated with the combinatorial curves of Lissajous. Trigonometric forms like Lissajous curves had already been used for light drawing in the mechanized slides of the nineteenth-century magic lantern, itself an illusionistic medium. The magic lantern was the direct forerunner of today's lamp projectors, a simple light source that illuminated and projected translucent slides onto walls or screens. These slides could be photographs, still images painted on glass, or more unusually, mechanical contrivances which produced mesmerizing motion effects. Mechanical slides, turned by simple gears, generated parametric combinations of abstract patterns, blown up by projection to a sublime scale. Among the numerous mechanical magic lantern slides, including those used to create moiré effects or chromatic oscillations, one of the most remarkable was the cycloidotrope.[55] This variant of Giambattista Suardi's geometrical pen traced cycloidic curves on a sooted plate as the magic lantern projected the drawing to cinematic proportions. Light drawing thus emerged not as a device of exact accounting but rather as an engine of endless wonder.

When overlapped and underlaid, Lissajous figures accumulated into complex interference and moiré patterns similar to *guilloché*. That interference suggested an intricate illusion of space, a parallax of volumetric figures that were an incidental consequence of highly mathematized procedures. Such striking effects inspired experimentalists to build elaborate devices to explore the visual gamut of Lissajous figures. Among the most ambitious of these machines was William F. Rigge's 1927 device for combinatorial curve drawing.[56] Through the permutation of gears in various sizes, periods, and configurations, it could construct cycloids, harmonic curves, and exact Lissajous curves, including stereoscopic plates of such figures. Rigge, an astronomer and reverend, spent over a decade perfecting this machine, expanding the parametric variation and combinatorial range of the device to billions of distinct harmonic curves—7,618,782,490, to be exact.[57] The machine thus became an empirical means of exploring an almost unlimited space of parametric permutations of curious mathematical volumes. In a particularly suggestive diagram, published alongside other research in Rigge's book *Harmonic Curves* (1927), a matrix of curve families and their continuous intertransformations emphasizes the arresting stereoscopic depth effects that were possible. Rigge's drawings were virtual objects whose implied surfaces were registered by the self-intersections of single continuous curves. Suggesting that the curves described are actually spatial constructs, Rigge devoted considerable effort to describing how his machine could generate stereograms of curved forms, which the user could then experience directly with a stereoscope. These stereograms depicted space curves that had never been constructed (or even calculated) in three dimensions but instead were represented in mutually coordinated partial projections that produced the illusion of a virtual volume. Rigge thus brought an immersive aspect to virtual volumes, connecting a technology of calculation to a technology of spatial perception.

As electronic machine control advanced in the mid-twentieth century, new devices that unified perception and drawing in a single cybernetic loop opened generative spaces for design. The experimentation with kinetic form-making machinery intensified after World War II, a time of exploding electrotechnological innovation and cultural fascination with scientific artifacts. World War II introduced a new generation of digital and analog electronic calculating machines, including the M5 and M7 ballistics computers used for measuring the trajectories of bombs and missiles. With these devices, the graphic methods of physics were not merely documentary but predictive. Conceived as advanced viewfinders with attached computers, they calculated the trajectories of bombs in real time while integrating wind, motion, and weather conditions. These bombsight computers were combinations of amplified drawing and augmented seeing machines. When bombsight computers were discarded after the war, inventive artists—many of whom had been trained in combat on these devices, and who thus internalized a mechanical intuition of their function—suddenly had access to powerful machines that enabled unprecedented visual effects.

The results of this cybernetic unification of perception and drawing were vertiginous images of overlapping and intersecting Lissajous-like curves. Among the most virtuosic of those exploring the bombsight computer as a medium of new drawing was the British machine experimentalist Desmond Paul Henry (1921–2004). By attaching missile guidance systems to combinatorial pen mechanisms inspired by Giambattista Suardi's eighteenth-century drawing machines, Henry generated curves that he described as "densely packed helices subject to various degrees of distortion, both intrinsically and extrinsically."[58] For Henry, the drawings were biological and almost alive: he linked his drawings to biologist D'Arcy Thompson's formal analysis of spicules, cells, and cellular aggregations. His machines synthesized the evolution of biomorphic and mathematical structures. Their tactic of interweaving kinetic tracery suggested that these drawings were not mere two-dimensional graphics but instead projections of shrouded bodies. Henry's drawings exhibited a vital asymmetry, with cloudlike imperfections and accentuated aberrations that he described as an "aleatoric note." In these aberrations one sees the trace of both the precision and the stochastic reality of the underlying physics.

figs. 6.12, 13 In their development of ever more complex machinery for the drawing of Lissajous virtual volumes, experimentalists also expanded the scale of the machines themselves. The German architectural photographer Heinrich Heidersberger (1906–2006) was among the most prolific of these experimentalists. Heidersberger's photographic work, which included modern architecture as well as images of Germany's industrial rebirth after World War II, revealed a predisposition for curved structures that hinted at the forms of his more generative work. Beginning in 1955, he undertook the construction of an elaborate device for directly tracing a vibrating and oscillating light beam on the photographic plate itself. In a hardware hacking process that recalled Desmond Paul Henry's ingenious machine, over the course of the next decades, Heidersberger refined several distinct iterations of an instrument he called the "rhythmograph." The product of this machine was a series of "rhythmograms," luminous images of automatically generated bundles of curves. These rhythmograms had deliberately spatial properties, engaged a relativistic sense of space and time, and captured some of the cinematic and combinatorial possibilities latent in light drawing.

Graphing Physics, Mapping Space

In 1952, the American physicist Donald Glaser (1926–2013) introduced a strange new kind of physical graphic apparatus: the bubble chamber. The heir apparent to the cloud chamber's diaphanous diagrams of subatomic particles and cosmic rays, the bubble chamber offered a new glimpse into the unseen spaces of fundamental physics. The images produced in bubble chambers were fields of cascading spicules, spirals, and helices that revealed a new logic of reality. Shimmering and scintillating, their forms had an enigmatic complexity. Yet they were genetically connected to Marey's motion studies, to Moholy-Nagy's virtual volumes, and to the myriad of other light-drawing experiments over the previous and subsequent decades. By indexing physics, graphic methods offered a unifying visual language of process and space.

fig. 6.14

Light drawings confront the viewer with predicaments of precision and reference. As tools of scientific experiment, they were used by physicists to try to overcome ambiguity and vagueness with incontrovertible evidence. In this capacity, light drawings were intended as narrow and indirect replicas of a certain state of concealed space. Yet, even graphically, they were never as precise as that idealization: auras, halos, flares, and other artifacts of light could impart the turbulent appearance of an obscuring fog or a churning storm. Moreover, there was a slipperiness about what should hold ontological and epistemic priority in the conversation between object and image: was the invisible entity the true reality,

6.12 Heinrich Heidersberger with a version of his rhythmogram apparatus.
Source: Institut Heidersberger. © Heinrich Heidersberger, Wolfsburg.

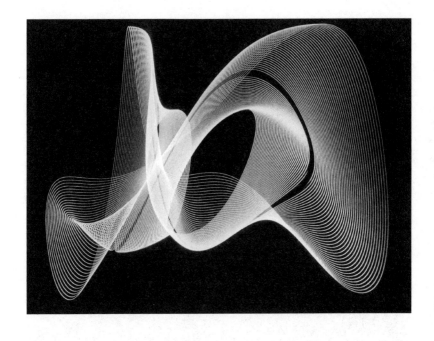

6.13   An example of the light drawings generated by Heidersberger's rhythmogram device.
Source: Institut Heidersberger. © Heinrich Heidersberger, Wolfsburg.

6.14   Image from bubble chamber at Brookhaven National Laboratory. Photograph taken
March 12, 1964.   Source: "Cascade 0 Production in 14-Inch Bubble Chamber at Brookhaven
National Laboratory," 1964, photograph, *Photographs Documenting Scientists, Special Events, and Nuclear Research Facilities, Instruments, and Projects at the Berkeley Lab, 1996–2014.*
© 2010–2019 The Regents of the University of California, Lawrence Berkeley National Laboratory.

or rather its visible trace? The dyad of object and image was locked together in an essential but undecidable epistemic dilemma.

Inscribing mathematical curves from physical phenomena was an encoding of reality that could unify and synchronize visual imagination and technological precision. Hungarian designer and Moholy-Nagy protégé György Kepes (1906–2001) gave voice to that aspiration: "With instruments we are gradually finding a common denominator in all sensed experience; it is possible to convert sound to sight, space to time, light to form, and interchange phases and events, static and dynamic, sensible and conceptual."[59]

In his funerary eulogy for Moholy-Nagy, Walter Gropius declared: "His greatest effort as an artist was devoted to the conquest of space, and he commanded his genius to venture into all realms of space . . . he constantly strove to interpret space in its relationship to time, that is, motion in space."[60] Moholy-Nagy's vision of space drawing in light was extended and transformed by subsequent designers in ways that would have been impossible to anticipate but which were in tune with a culture striving for a unity of technology and experience. The mechanically drawn chronophotograph became a means to that unity. With it came new ways of seeing, measuring, and imagining space.

Gropius's rational approach to illusion on the one hand and Moholy-Nagy's experimental fusion of light and machine on the other bounded a specific method of visual instrumentation which presaged the virtual space of light drawing. Partway between scientific measurement and imaginative fantasy, light drawing quantified irregular or unpredictable motion as an oscillating line of pure signal. The light-motion study as a way of seeing also became the nucleus of a subculture of exact but unexpected generative drawing. With the earliest motion studies, human and machine action became comparable, even perhaps interchangeable. Light became a pencil to draw the silhouette of behavior. What unified this subculture was a consistency of representation, an attraction to the pure faint phantom forms of traced light. Part illusion, part flickering vision of a luminous new realm, light drawings became faint sketches of an electromechanical future.

7 Labyrinths, Topology, and Meinong's Jungle: Architecture's Impossible Objects

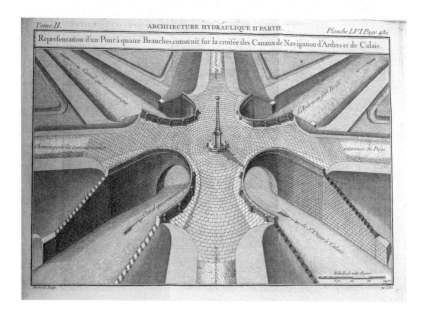

7.1 Jean Perronet's curiously topological proposal for a bridge at Pont Sainte-Maxence, near
Paris, 1774, intended to "bring together at one point all the principal communication routes of the
province." Source: Robin Evans Slide Collection, Architectural Association Archives, London.

In 1904 the Austrian philosopher Alexius Meinong (1853–1920) fig. 7.1 promulgated a theory of nonexistent objects, or objects which were, by their definitions, paradoxical. Meinong's nonsense entities included such rarities as the proverbial square circle,[1] the golden mountain, or the current king of France.[2] Meinong toyed with the linguistic function of reference: as the philosopher Dale Jacquette succinctly put it, "all thought intends an object, but not all intended objects exist."[3] More than a mere philosophical and logical diversion, the problem that Meinong articulated— whether impossible objects have some sort of conceptual existence, since they can be *referred* to without contradiction— scandalized philosophers like Bertrand Russell, who saw it as an attack on the validity of mathematics itself. Later theorists referred to the lush and fecund realm of referenced but impossible objects as Meinong's Jungle.

Drawings have a referential capacity analogous to language. As representations, drawings point to entities other than themselves: buildings that have been, are, or could be built. But that referential capacity introduces strange existential disjunctions for architects: some drawings may intentionally defy physical construction, referring to impossible spaces. Was there architecture in Meinong's Jungle? Certain twentieth-century architects were preoccupied with illusional drawings that seemed impossible in two dimensions but, through some mathematical sleight of hand, actually had Escher-like resolutions in three. As we have seen, Walter Gropius insisted that optical illusions—particularly those of projection and perspective—were essential training for the modern architect. In addition to energizing interest in representational ambiguity, modernism also affiliated architecture with a pared-down, quasi-mathematical drawing lexicon of linework wireframes. Through odd new drawings evoking objects caught between dimensions, intuitive dichotomies between front and back or inside and outside were undone. All parts of an object could be seen simultaneously, through strange contortions of surface and form. Topology was the mathematical language that brought a fantasy of cubist omniscience within reach.

Colloquially known as *analysis situs*, "the science of situation," or "rubber sheet geometry," mathematical topology delimits the qualitative spatial relationships of surfaces, patterns, and volumes that do not change if they are continuously stretched or warped. These invariant facts—for example, that one surface is joined to another along a specific edge or that a surface has a fixed number of apertures—persist even if the measured geometric dimensions of an object wildly contort. They enable the description of forms that would frustrate more standard orthographic drawing. Topological invariants like connectivity, intersection, or adjacency define deep characteristics of shape. Two entities are said to be homeomorphic or topologically equivalent if they share these invariant characteristics. To cite a classical example, though they differ geometrically, the sphere and the cube are topologically equivalent since they can each be continuously transformed into the other. Invariant topological characteristics establish what the philosopher of mathematics Brian Rotman has called "types or species or families of related objects to which they are structurally akin."[4] From the early twentieth century it was apparent that these topological families extended to the organizations and forms of virtually every artifact of architecture.

Topology untangles aberrant, anomalous, and slippery forms like Möbius strips or Klein bottles by furnishing an exact language for their description. Such geometric entities have the bizarre property that their front and back faces are identical: one could run a finger along the surface continuously, beginning on one side, only to find it on the other side. For architects who saw their potential, topological inversions became a practice by which perceptual paradox and spatial fact could be squared, endowing design with a radical optical and spatial pliancy. Since topology brought the mundane facts of building layout into the same language as exuberant flights of form, those two could be identified and become equally dynamic. Most profoundly, topology is a language to limn the deeper structures of spatial organization, laying bare the logical foundation of design's ancient quest for parts-to-whole relationships. A single building was no longer a discrete instance in a sporadic set of exemplars but rather one of a continuous family of elastic differential variations. Topology was a language of formal structure by which the whole of architecture might be decoded, compared, and classified.

fig. 7.2
Though topological ideas have been circulating in architecture for nearly a century, the interest in topology among the

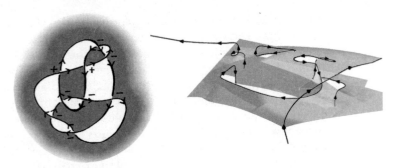

Seifert diagram basis of main geometrie

Reference Seifert surface

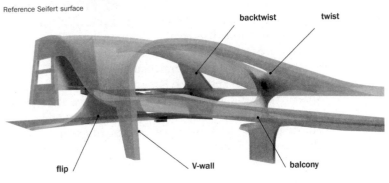

backtwist   twist

flip   V-wall   balcony

7.2   A matrix of analytic diagrams showing the genesis of Arnhem Station from a knot-generated Seifert surface. Courtesy UNStudio.

neo-avant-garde in the 1990s has long outshone earlier antecedents. In 1997, Ben van Berkel and Caroline Bos of UNStudio argued that "it is only in recent years that the space of liquidity, the directionality of topology, or the structure of viscosity have begun to assume a significance in architecture."[5] Knots, folds, and myriad distortions became vital tools in that moment of formal experimentation, producing building proposals that explored provocative topological forms. In their project for a central rail station in Arnhem, Netherlands, Van Berkel and Bos explained that "the topology of relations" catalyzed "the introduction of a diagram that encapsulates the technical/spatial organization."[6] Intriguingly, the organizing diagram was not an innocent sketch but rather the Seifert diagram, a specific mathematical technique whose strict application reveals a certain level of mathematical literacy and conceptual rigor. Using certain graphic notations, Seifert diagrams describe how a self-intersecting planar curve drawing should be interpreted as a three-dimensional non-self-intersecting surface. The Arnhem Seifert diagram evinces all the exact mathematical annotations required by knot theory—arrows indicating the directional sense of the curve, checkerboard solid/void alternation of partitioned regions, and specifications of how the curve path overlaps and underlaps itself. A particular and notably obscure mathematical technique was applied exactly to generate performative organizations of architecture. Through what they termed an "instrumentalizing technique" aided by digitization, Van Berkel and Bos saw a potent new practice of spacemaking in the manipulations of topology. Software that encapsulated and enabled topological operations—for sewing, stitching, and suturing surfaces—was a paradigmatic catalyst for these new frontiers. Digital topology could surpass the limits of orthogonal descriptive geometry to define forms that mere projective techniques could never fully draw.

Yet the obscure seeds of a topological architecture had actually been planted far earlier. Even in the 1930s certain architects recognized the essential power of topology, and saw not one topology but two: a topology of form and a topology of function. The first was the mathematics of pliant, elastic shapes, a mesmerizing corollary of differential geometry that took topology as an explicit spatial premise. The second topology was a cartography of operational networks like circulation pathways or communication webs that crisscrossed the plans of buildings to limn their logical structure. From the 1930s through the 1970s, these two distinct trajectories developed in parallel, each comprising the kernel of a distinct topologically oriented thread of architectural experiment. In its dual formal and functional guises, topology was a vehicle to probe the strange interstitial space between perception and spatial operation. Architecture could overcome apparent impossibilities—geometric, perceptual, operational—through the space-warping machinery of topological mathematics.

Dual Topologies I: Topologies of Form

The precursors of architectural topology lie in the peculiar visual culture of late nineteenth-century surface maquettes and *analysis situs* diagrams that coalesced around nascent mathematical knot theory. As these representations intersected and hybridized, abstract diagrams condensed into new mutations at the limits of spatial possibility. As historian of mathematics Moritz Epple has pointed out, knots—continuous closed curves that self-entwine in space—were a vital emerging topic not only in mathematics but also in chemistry and physics during the 1870s. Physicists such as James Clerk Maxwell and William Thomson (Lord Kelvin) suspected that the vortical geometries of knots could describe the deep structures of magnetic fields or even atomic interactions, unlocking the benthic mysteries of reality itself.[7]

The entwined work of two Scotsmen—the mathematician Peter Guthrie Tait (1831–1901) and chemist Alexander Crum Brown (1838–1922)—moved these theoretical relationships toward more visual taxonomic diagrams and physical constructs. Tait ventured the first rigorous classifications of complex knots, proposing measures of "beknottedness" and compiling tables of knot classes in his "orders of knottiness."[8] Tait's elegant two-dimensional diagrams were beguiling in their own right but only hinted at the lush three-dimensional spaces that were actually implicated in spatial knots. Inspired by his brother-in-law Tait, Crum Brown recast knot diagrams as fully spatial objects in his absorbing 1885 paper "On a Case of Interlacing Surfaces."[9] In it he proposed endless tapestries of intricate twisted membranes arranged in

fig. 7.3

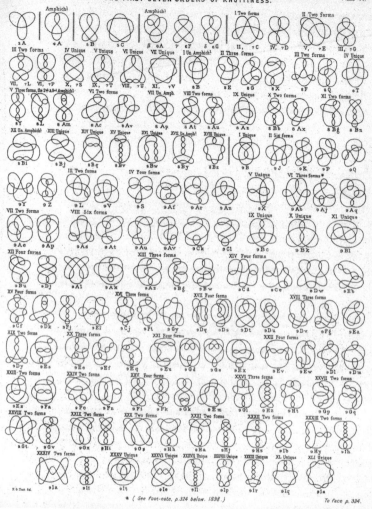

7.3 Peter Guthrie Tait's "First Seven Orders of Knottiness," one of the earliest topological classification tables. Source: Peter Guthrie Tait, *Scientific Papers* (Cambridge: Cambridge University Press, 1898), pl. VI.

repeating lattice structures, generated from knotlike diagrams. Crum Brown brought a keen attention to the physical artifact of an interwoven surface as opposed to the more abstract linear filaments of Tait's tables. A certain spatial intuition was at work in his curious braided surfaces, which were akin to taut membranes spanning intricate wire edges. These mesmerizing surface-lattice oscillations were the wellspring of entirely new types of endlessly fluid surface patterns. Weaving—not only as a theoretical analogy but also as a tactic of surfacial construction—became a practice through which Crum Brown could explore topological inversions and twisted labyrinths. The entrancing spatial complexities of Crum Brown's physical models seem to shimmer between dimensions, half diagram and half sculpture.[10]

fig. 7.4

Though knot theory made them possible to conceive, Crum Brown's models were still exceedingly daunting to construct. Spanning torqued and twisted loops was a complex geometric puzzle in its own right. The newly discovered minimal-surface films of the Belgian physicist Joseph Plateau (1801–1883) unraveled this puzzle by offering a miraculous, almost instantaneous way to build—or at least fleetingly visualize—fantastically complex knot surfaces in thin glycerin films. In fact, Tait had explicitly endorsed Plateau's remarkable discoveries in his early development of knot theory.[11]

figs. 7.5, 6

Plateau's key discovery was the minimal surface: a glycerin film that naturally assumed an anticlastic and area-minimizing shape in perfect uniform surface tension, at uneasy but arresting equilibrium.[12] These exquisite thin-film membranes, only molecules thick, were produced from wire armatures submerged in a specially formulated solution similar to common soap.[13] Though his interests concerned the physical limits of liquid surface tension, in his 1873 *Statique expérimentale et théorique des liquides soumis aux seules forces moléculaires* Plateau used this method to produce a menagerie of complex geometric forms. Because they might burst at any moment, Plateau's films were immediately documented by the photographer Adolphe Neyt in iridescent stereographs.[14] Plateau's shimmering, transparent film veneers were both miniature tensile structures and evanescent mathematical models. Using his minimal surfaces, exotic classes of impossible-seeming objects materialized in quotidian physical space.

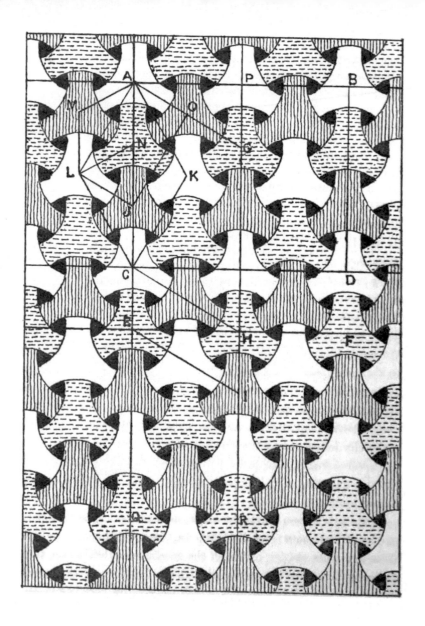

7.4 Alexander Crum Brown's diagram of a lattice of interlacing knot surfaces.
Source: Alexander Crum Brown, "On a Case of Interlacing Surfaces," *Proceedings of the Royal Society of Edinburgh* 13 (1885–1886): 382–386.

Until the early twentieth century, architects were oblivious to topological innovations, and architecture took for granted dichotomous distinctions like up and down, inside and outside, or left and right. But around the 1930s, topology became a vital concern for designers who ventured serendipitously into the realm of warped shapes. Tentative at first, these early forays informed a lasting intuition for what topology was and how it could further the ends of design. The Swiss architect and educator Max Bill (1908–1994) was among the earliest to consider topological ideas:

> In the winter of 1935–1936 I was assembling the Swiss contribution to the Milan Triennale, and there I was able to set up three sculptures to characterize and accentuate the individuality of the three sections of the exhibit. One of these was the "Endless Ribbon," which I thought I had invented myself. It wasn't long before someone congratulated me on my fresh and original reinterpretation of the Egyptian symbol of infinity and of the Moebius ribbon. I had never heard of either. My mathematical knowledge had never gone beyond ordinary architectural calculations, and my interest in mathematics was not very great.[15]

Evidently shocked by this development, Bill let topological questions lay fallow for a decade until he realized the Möbius band was not an isolated pronouncement but the first word of an expansive new design vocabulary. By the 1940s, Bill had fully embraced topology as a catalyst of creative form making and as an element of the "mathematical line of approach" to the arts that he sought.[16] Bill recounts: "ever since [the] 1940s I had been thinking about the problems of topology. From them I developed a kind of logic of form."[17] In temporary sculptures like the 1947 *Kontinuität*, and in more permanent pieces such as *Rhythmus und Raum* of 1948, Bill probed a newly elastic idiom of spatial enclosure. The spatial vocabulary of nineteenth-century knotted surfaces provided a strange new visual paradigm for twentieth-century design.

While designers like Bill intuitively explored new topological forms, in the early twentieth century mathematicians found themselves in closer conversation with designers, often through little more than accidents of circumstance. Among these

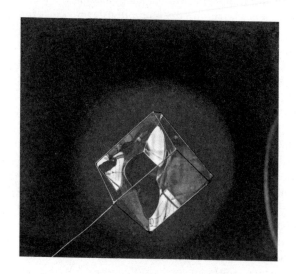

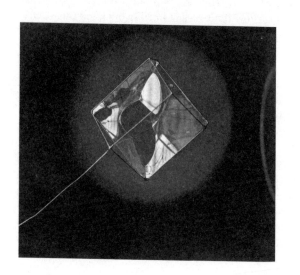

7.5   A stereoscopic view of one of Joseph Plateau's minimal-surface soap films, 1873.
Source: Joseph Plateau collection, Ghent University Museum.
Courtesy of GUM—Ghent University Museum.

7.6   Wire armatures used by Plateau to develop minimal-surface soap films.
Source: Joseph Plateau collection, Ghent University Museum.
Courtesy of GUM—Ghent University Museum.

unexpected interlocutors was German mathematician Max Dehn (1878–1952), a student of the imminent mathematician David Hilbert and a renowned topologist in his own right. Hilbert introduced the public to a visuality of topology in his phenomenally popular book *Geometry and the Imagination*,[18] and perhaps unsurprisingly for a student of Hilbert and a product of Felix Klein's Göttingen school, Dehn laid critical foundations for topology with a refined geometric intuition. In 1907, he and the Danish mathematician Poul Heegaard stated the first rigorous version of the classification theorem for orientable and nonorientable surfaces, which described an exhaustive way to conceptually catalog every possible (and impossible) surface for the first time.[19] His work was not only taxonomic but also generative: the eponymous process known as Dehn Surgery furnished an algorithm by which to combinatorially construct intricate topological surfaces by cutting, warping, and restitching existing surfaces to each other. Variants of Dehn's cut-and-sew combinatorics would prove essential to later designers tackling the arcane geometry of knot surfaces.

Perhaps most importantly for architects, Dehn was a pioneer in teaching complex ideas of topology to designers and artists. His hybrid pedagogy—part mathematics, part design—cast topology as vitally relevant to aesthetic questions. The mathematician Philip Ording recounts how after immigrating to the United States, Dehn joined the faculty of the nascent but critically influential Black Mountain College in 1944 as the only mathematician among a rotating cadre of designers, artists, architects, and composers that included the composer John Cage, the choreographer Merce Cunningham, and the architect Buckminster Fuller.[20] He lectured on the history of mathematics and its encounters with ornament and the arts.[21] In his course Geometry for Artists, Dehn introduced geometric constructions from elementary conic reflection lines and curve tangent fields to more sophisticated polyhedral bodies and applications of Desargues's theorem. The drawing archive of Dehn's students reveals surprisingly advanced constructions of conic and hyperbolic geometry.[22] Dehn also taught key topological invariants like the Euler characteristic, which posits an invariant arithmetic relationship between a polyhedron's faces, edges, and vertices. At Black Mountain, he opened exchanges with the major designers who taught there, including Buckminster

Fuller and Josef Albers.[23] Concepts once confined to the closed culture of elite mathematics were shifting into formal and intellectual adjacency to design.

The Gestalt of Impossible Objects

Through a mélange of happenstance, curiosity, and focused intention, topological mathematics gradually infiltrated the culture of design. But ironically it was not mathematics per se but the conundrums of gestalt perception that provoked designers to engage topology in a sustained way. Gestalt theories posed basic questions at the nexus of visual experience and object recognition that confronted the possibility and impossibility of perception itself. What was the minimum graphic information required to represent a three-dimensional object? When does the perception of one object bifurcate in two, or two fuse into one? Facing such questions, topology was tacitly implicated in the experimental psychology of perceived form. The Austro-Hungarian psychologist Max Wertheimer (1880–1943), one of the earliest developers of so-called Gestalt theories, grappled with such enigmas in his 1912 *Experimentelle Studien über das Sehen von Bewegung* (Experimental studies on the vision of motion) and in *Untersuchungen zur Lehre von der Gestalt* (partially translated as *Laws of Organization in Perceptual Forms*).[24] Particularly in *Laws of Organization in Perceptual Forms*, Wertheimer employed elementary graphic diagrams—point grids, simple curves, and basic polygon intersections reminiscent of Tait's knot taxonomies—as tests to interrogate the thresholds of ambiguous perception and the relationship of parts to whole. Wertheimer argued that the fusion or distinction of figures was contingent on the relationship of their perceived boundaries: "It is instructive in this connection to determine the conditions under which two figures will appear as two independent figures, and those under which they will combine to yield an entirely different (single) figure. . . . By means of what additions can one so alter the figure that a spontaneous apprehension of the original would be impossible?"[25]

The visual puzzles of the gestalt experiment uncovered odd spatial paradoxes in simple two-dimensional diagrams. Are the objects implied by interlaced figures transparent, overlapping, or

fig. 7.7

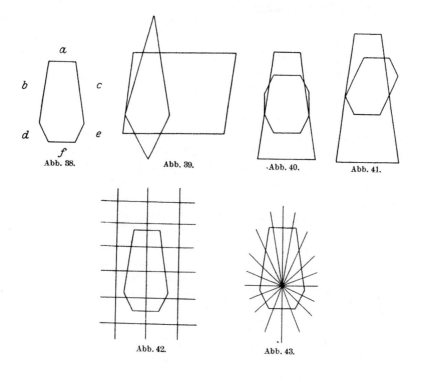

a
b c
d e
f

Abb. 38. Abb. 39. ·Abb. 40. Abb. 41.

Abb. 42. Abb. 43.

7.7 Diagrams from Max Wertheimer's "Investigations on Gestalt Principles" showing the coherence and incoherence between overlapping figures.   Source: Max Wertheimer, "Untersuchungen zur Lehre von der Gestalt, II," *Psychologische Forschung* 4 (1923): 328.

knotted together? Gestalt theory claimed to unravel these mysteries, and designers found its arcane games irresistible. Lectures on gestalt vision presented at the Bauhaus in 1928, and 1930–1931 were attended by faculty like Paul Klee, Josef Albers, and Wassily Kandinsky.[26] Klee was particularly enamored of Wertheimer's notion of figural integrity and the boundary between integral and disintegrated form, importing gestalt figural exercises into his Bauhaus courses.[27]

Notwithstanding Klee's interest, it was his student the German artist and designer Josef Albers (1888–1976) who most decisively transmitted gestalt and topological inclinations to later sculptural and architectural designers. Albers's *Structural Constellation* series, a collection of drawings, prints, and lithographs, tacitly explored themes of gestalt coherence, projection, and topology hovering at the nexus between figural diagram and spatial proposition. Produced between 1937 and 1960, the series' initial appearance roughly coincided with Albers's time on the faculty of Black Mountain College. In these pieces, collections of closed polylines that overlap, underlap, and transparently interlace are arranged as if to evoke paradoxical axonometric spaces. Art historian Anthony Auerbach observes that "their forms are rigorously defined, yet disintegrate in a multitude of perceptual possibilities and contradictions."[28] Projective ambiguity begets a topological ambiguity. Which polygons imply the edges of surfaces, and which are merely transparent wireframes? One wishes to dip them in glycerin to instantly construct one of Plateau's illuminating models. Yet the impossibility of so simply untangling the *Constellations* is precisely their inscrutable attraction. The misleading clarity of the drawing is subverted by the multiplicity of possible readings, suggesting not one but many tumbling and nonorientable volumes. The *Constellations* recall the duality of reversible images or the pseudoscopic phenomenon of the Necker cube.[29]

fig. 7.8

Albers's *Constellations* captured continuous oscillations between figural fusion and dissolution. Their meticulous craft sponsored that illusional shimmering. They were painstakingly constructed on graph paper before their transfer to pin-pricked unruled paper, creating what Auerbach has described as "astral maps."[30] This graph paper, far from being a marginal detail,

7.8   Josef Albers, *Structural Constellation*, ca. 1950.
Source: Josef and Anni Albers Foundation, 1976.8.1715. © The Josef and Anni Albers Foundation /
Artists Rights Society (ARS), New York, 2020.

enforced a crystalline structure on the constellations, allowing them the possibility of infinite replication and patterned extension akin to Crum Brown's interlaced lattices. In effect, the graph paper embedded the surfaces in a crystal frame. While the *Structural Constellations* feel almost deductively inscribed from a priori principles, Albers's inclination was immediately experiential. He wrote that "no theory of composition by itself leads to the production of music, or of art.... What counts here—first and last—is not so-called knowledge of so-called facts, but vision—seeing. Seeing here implies Schauen (as in Weltanschauung) and is coupled with fantasy, imagination."[31] For Albers, the experiments represented by the *Constellations* exceeded the medium of representation to address visual imagination directly, and thereby establish an intuitive rapport between sight and geometry. Obscure ciphers of phantom realms, the *Constellations* peer into quasi-spaces on the boundary of fantasy and fact.

Albers wrote very little about the *Constellations*, and so any ascription of intellectual motives must necessarily be indirect. Yet there are clues in his personal library, which includes books on the problems of gestalt psychology,[32] the intersections of mathematics and art,[33] mathematical popularizations such as Hermann Weyl's famous *Symmetry*,[34] and even works by some seminal figures in knot theory.[35] While Albers may not have engaged these topics per se, he was no stranger to the theorizations of perception and mathematics as they related to visual culture.[36]

Architectural Knots

The graphic and quasi-spatial manipulations of Albers's *Structural Constellations* recall not only Tait's knot catalogs, but even more the very particular knot theory of German mathematician Herbert Seifert (1907–1996), developed in the 1930s. Seifert's diagrams had a remarkable spatial affinity to Albers's sensibility, to the methods of his students, and ultimately to the revival of architectural topology in the 1990s. In 1935 Seifert published "Über das Geschlecht von Knoten" (On the types of knots), in which he described classes of complexly curving surfaces constructed to have a specific space curve as a boundary.[37] Seifert ingeniously resolved the paradox of two-dimensional intersection points by identifying them as

fig. 7.9

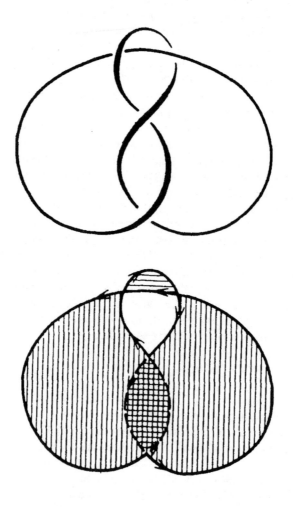

7.9   Seifert's diagrams of surface partition parity.
Source: Herbert Seifert, "Über das Geschlecht von Knoten," *Mathematische Annalen* 110 (1935): 572.
© 1935 Springer Nature.

locations where a three-dimensional surface twists on itself. He introduced algorithms, diagrams, and nomenclature for constructing non-self-intersecting three-dimensional surfaces from self-intersecting two-dimensional knots, thus resolving their impossible spatial implications. It was exactly these diagrams that Van Berkel and Bos employed in their Arnhem station sixty years later. Like Möbius strips or minimal surfaces taken to the ultimate degree, the so-called Seifert surfaces transform chains of curves which appear to overlap, intersect, or interweave into single continuous surfaces. They are entities which are impossible to fully comprehend in two dimensions, and yet unfurl integrally in three. In effect, they provided a mathematical answer to Wertheimer's gestalt conundrums, an algorithm for constructing objects that simultaneously validated and defied perceptual conventions.

Seifert surfaces and *Structural Constellations* exemplify objects which, when delineated by a single two-dimensional drawing, appear spatially inconsistent or even paradoxical. Fusing interior and exterior and frustrating orientation, these drawings often adopt a shimmering and reversible axonometric presentation. Albers's drawings, for example, hover at a curious intersection of vision, psychology, and mathematics, inviting plural interpretations and confounding the apprehension of their implied space. Yet their three-dimensional counterparts intimated by Seifert were even stranger, a terra incognita of forms that aimed to amplify the productive complexities of dimensional ambiguity.

The fantastic sculptural and architectural projects of Albers's intellectual descendants extended this research of integral and continuous form along progressively more spatial lines. By the 1950s, topological sculpture, encouraged substantially by Albers, emerged as a fertile terrain between gestalt enigma and architecture. Through Seifert-like patterns propagated across modular and crystalline lattices, designers multiplied the delicate effects of light and enclosure afforded by knot forms. Albers was a nucleus and sponsor of such research. Since his time at the Bauhaus, and later at Black Mountain College (1933–1949) and Yale (1950–1958), he taught methods of structural sculpture, including intricate paper folding. While Albers himself never developed a fully three-dimensional sculptural analogue to his *Structural Constellations*, he nurtured and promoted the work of artists and designers who did exactly that, bringing impossible objects from the page into the physical world. In 1961, New York's Galerie La Chalette convened a landmark show entitled *Structural Sculpture*. The show gathered seven of Albers's students including Robert Engman (b. 1927)[38] and Erwin Hauer (1926–2017), whose work provided three-dimensional answers to the implied topologies of Albers's drawings.[39] In the catalog of the show, Albers described work toggling between plane and volume:

> It is more than 40 years since Gestalt psychology discovered, and proved, that three-dimensionality is easier and earlier perceived than two-dimensionality. This means it is more difficult to recognize two dimensions than three dimensions. . . . But finally a few independent [designers]—only a few in sculpture, as a few in architecture—were courageous enough to concentrate on the plane—the in-between of volume and line—as a broad sculptural concept and promise. It is a promise truly new and exciting . . . because it traverses continually the separation of 2 and 3 dimensions.[40]

In an echo of the hyperdimensional forms of Theo van Doesburg a generation before, this cohort of designers now searched for shape in the cracks between dimensions.

figs. 7.10, 11

Structural Sculpture offered weird new bodies of seamless continuity. Of the several contributions to the show, the dizzying masonry labyrinths of Erwin Hauer extended Seifert-like surfaces most decisively toward architectural epiphany.[41] Porous yet substantial, they made a web of a thousand Möbius strips, beckoning the eye to endlessly weave in and out of their absorbing microarchitecture. In fact, these were not mere walls but were all the elements of architecture—doors, windows, passages, canopies—melted together in a continuous frieze. Hauer applied a rigorous but intuitive language of depth, overlap, and occlusion in torqueing volumes, and was inspired in part by the abstract work of the British sculptor Henry Moore (1898–1986). During the late 1930s Moore had himself mined Fabre de Lagrange's mathematical models at London's Science Museum for spatial ideas, and Hauer was a second-generation heir of the sensibility of these Belle Époque mathematical maquettes.[42] Hauer's paradigmatic works

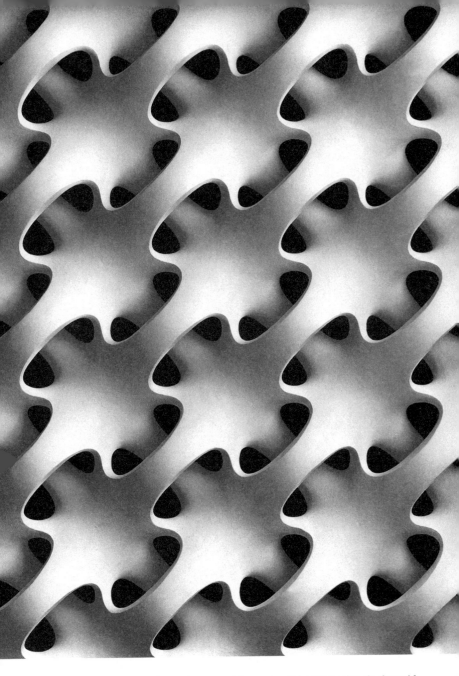

7.10 Erwin Hauer, *Continua Design #1*, a modular constructivist deformation of a plane with
perforations, developed in 1950 and deployed as a wall structure.
Courtesy of Princeton Architectural Press.

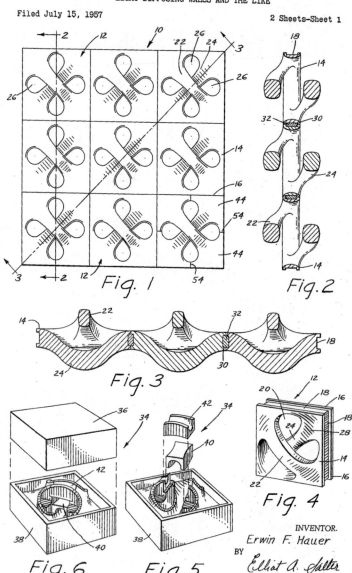

May 21, 1963   E. F. HAUER   3,090,163

LIGHT DIFFUSING WALLS AND THE LIKE

Filed July 15, 1957   2 Sheets-Sheet 1

Fig. 1

Fig.2

Fig. 3

Fig. 4

Fig. 6

Fig. 5

INVENTOR.
Erwin F. Hauer
BY
Elliot A. Salter

7.11   One of Hauer's several patents.
Source: Erwin Hauer, US Patent 3,090,163, filed July 15, 1957, and issued May 21, 1963.

were his monumentally intricate walls composed of unitized cast knots arrayed as continuous tapestries, reminiscent of Crum Brown's interlaced surfaces but rendered in curvilinear concrete. Hauer had already constructed several of his porous walls before moving from Austria to the United States, where he filed several US patents for "light diffusing walls" and extended his work to an architectural scale. Hauer advertised his "sculptural pierced walls" commercially as ready-made products in *Architectural Record* and *Progressive Architecture* throughout the 1960s, claiming to "offer to the eye of the observer a magical union of light and rhythmic forms."[43] Constructed in gypsum or concrete, his walls became both monuments and commodities of new topological form. They proved, at least at the level of ornament, that strange knot surfaces could indeed become very real elements of architecture.

Hauer augmented his methods by communicating with mathematicians whose research intersected with his own. In particular Hauer corresponded with the American mathematician Alan Schoen (b. 1924), who was at that time compiling his 1970 report *Infinite Periodic Minimal Surfaces without Self-intersections*, a breathtaking visual catalog of torqued surfaces that infinitely extend in the x, y, and z axes.[44] Like Crum Brown, Schoen studied lattice structures as the genesis for complex labyrinths. To aid his theoretical labors, he fashioned dozens of physical models and, extending Plateau's techniques, pioneered emerging computer modeling technology to generate surface wireframes stereoscopically. Schoen's and Hauer's work was so serendipitously convergent that Hauer claimed they had independently discovered what is now known to mathematicians as the Schoen surface.[45] Regardless of its true provenance, it was clear that the worlds of theoretical geometry and spatial design were inexorably intertwining.

fig. 7.12

By moving beyond discrete geometry toward innumerable but continuously equivalent in-between shapes, topology annexed a broad new territory of twisted forms that bordered on the illusional. Topology also articulated more precisely the graphic format in which rational form could be diagrammed, effectively replacing the constructed drawing with a sketch of abstract spatial and geometric relations. Diagrams could leap

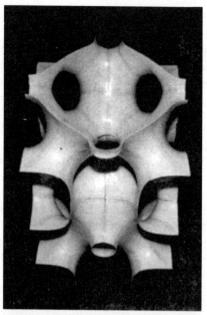

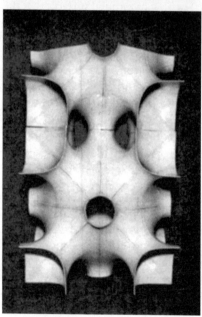

7.12   Models from Alan Schoen's notable catalog of triply periodic space-filling minimal surfaces, 1970.   Source: Alan H. Schoen, *Infinite Periodic Minimal Surfaces without Self-Intersections* (Washington, DC: NASA, 1970), 30.

directly to fantastically complex spaces, without the heavy intermediary of plan or section drawings.

A new optimism energized designers who glimpsed the possibilities of ever more ambitious applications for topological and minimal surfaces. The minimal-surface techniques of Plateau became not only spatial wish images but technical models for topological forms. Undoubtedly the best known and most influential architect to take on Plateau's films as architectural catalyst was the German designer Frei Otto (1925–2015), who built an entire research program around what Daniela Fabricius calls a "translation between incalculable materiality and calculable information."[46] Otto's work has been thoroughly considered elsewhere, but there were others who also embraced the topological project of minimal surfaces to great effect. One of the most daring instances was the development of the lyrically arching Ponte sul Basento by the Italian engineer Sergio Musmeci (1926–1981). This bridge, designed and built between 1967 and 1976, revealed the elegant minimalism possible with spanning knot surfaces. Because these spanning surfaces were derived from a solution with continuous surface tension, the bridge demonstrates that such surfaces can be adapted to act in compression as well. Mirroring Plateau's early minimal-surface experiments, Musmeci began with a schematic glycerin film geometry that he refined through relentless empirical testing of smaller-scale Perspex maquettes as well as much larger concrete load-testing models. The final structure, cast as one continuous body, reads as a scaled up excerpt from one of Hauer's knot-walls. Musmeci's bridge affirmed the performance of apparently impossible geometries of topological surfaces outside of a purely aesthetic or conceptual value. On the contrary, topological forms now pointed toward robust new structures, enabled by once-obscure machinery at the heart of modern geometry.

figs. 7.13–15

Topological Pedagogies

By the mid-1960s mathematical topology rewired not only the exceptional work of individual designers but also the curricula of entire design schools, bringing the spatial categories of design itself into question. Far from remaining an obscure and arcane mathematical exercise, topology was fast becoming a technology

7.13   Ponte sul Basento, Potenza, Basilicata, Italy. Minimal-surface soap film model.
MAXXI, National Museum of the Arts of the XXI Century, Rome. MAXXI Architettura Collection,
Archive MUSMECI Sergio—ZANINI Zenaide.

7.14   Ponte sul Basento, Potenza, Basilicata, Italy. Construction photograph.
MAXXI, National Museum of the Arts of the XXI Century, Rome. MAXXI Architettura Collection,
Archive MUSMECI Sergio—ZANINI Zenaide.

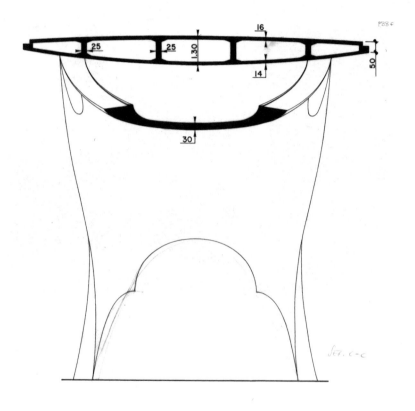

7.15 Ponte sul Basento, Potenza, Basilicata, Italy, section drawing.
MAXXI, National Museum of the Arts of the XXI Century, Rome. MAXXI Architettura Collection,
Archive MUSMECI Sergio—ZANINI Zenaide.

to solve unsolvable formal problems. Topological notions infused the whole pedagogy of Hochschule für Gestaltung Ulm (Ulm HfG), the progressive design school founded in Ulm, Germany, in 1953 by Inge Scholl, Otl Aicher, and Max Bill. The range of mathematical concepts pressed into the service of design at Ulm was staggering, ranging across fractal geometry (through the Sierpinski triangle),[47] combinatorics, curve theory, set theory, probability, game theory, linear programming, and information theory.[48] Intended to better equip the designer to influence an increasingly technological culture, these concepts were also co-opted as media of formal design research. In Ulm's architecture curriculum, the heroes of canonical modernism were jettisoned in favor of studies of modular construction underpinned by a strong theoretical interest in topology. Tomás Maldonado, the designer who led the school after Bill departed, envisioned an expansive role for topological methods, a role that could restructure the process of design itself.[49]

figs. 7.16, 17

Maldonado saw a precedent for his topological interests in the arguments of the British crystallographer J. D. Bernal, who recognized the potential of topological architecture as early as the 1930s. But Maldonado also expanded on Bernal's views to lay out a deeper vision of topological design:

> Bernal was one of the first to draw attention to the future importance of topology for architecture and town and regional planning. (In this case the combinatorical or algebraic form of topology is meant, and not the set-theoretic general topology.) Bernal's prediction has been confirmed, at least in one of them: the theory of linear graphs possesses a considerable instrumental value in the design of buildings where extremely complex circulation problems must be solved—hospitals, airports, stadiums, factories, theatres, and exhibitions. . . . Flexible architecture, which has been hailed so often, could become a reality if this architecture of transformations could be realized with the aid of the geometry of transformations, i.e. if this "rubber" architecture would be expressed in a "rubber" geometry. But that is a hypothesis, the validity of which must first be proved—in other words a hypothesis which cannot be confirmed yet.[50]

251

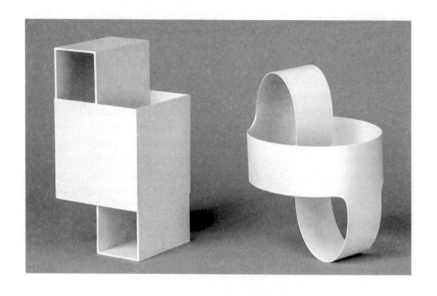

7.16   Nonorientable surface model exercises, Ulm HfG, which use a variant of topological surgery to construct twisting volumes. Lecturer: Gui Bonsiepe, study year 1965/66, department product design, HfG-Archiv / Museum Ulm (Inv. Nr.: D 1.0302, M 066, D 1.0304).

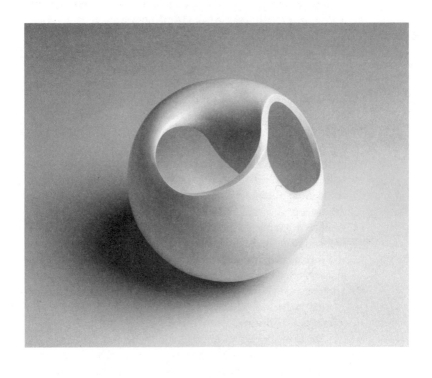

7.17   Nonorientable surface model exercise. Lecturer: Tomás Maldonado, study year 1956/57,
Student: Ulrich Burandt. Photographer: Ernst Fesseler, Bad Waldsee, 2003
© HfG-Archiv / Museum Ulm.

The functional topology of network structures and the formal topology of shape deformations appeared increasingly fundamental to decoding the deep structures of design problems. At that propitious moment, each was poised for significant advancements in design, including those from Ulm's own faculty.

figs. 7.18, 19 Ulm HfG's institutional project in topological design was buttressed by the research of individual faculty like Maldonado, and particularly the designer Walter Zeischegg. Zeischegg shaped Ulm both intellectually and administratively, teaching and serving on the school's executive committee for the duration of its existence.[51] His courses excavated the structured spatial transformations underlying topological surfaces.[52] In surprising synchronization with Erwin Hauer, Zeischegg extended the logic of periodic structure to the architectural scale of modular wall-knots. Indeed, Zeischegg and Hauer were resident in Vienna at the same time, and began almost simultaneously to work on the same problem of periodic space tilings. Zeischegg attended Vienna's Academy of Fine Arts and first began investigating these shells in 1947–1948, and exhibited them at the Academy in 1949 and 1950. Hauer attended the Academy for Applied Arts and also first exhibited his designs in 1950. Although Zeischegg was nearly a decade older, it is difficult to escape the possibility that the two crossed paths, and that postwar Vienna was an early crucible of topological sculpture.

In Ulm courses taught by Maldonado and Zeischegg, the two types of topology—formal surface geometry on the one hand and functional circulation networks on the other—were interwoven with a resolved commitment to mathematical diction and method. On the one hand, Dehn's surface surgery appeared in the constructed diagrams of topological surface exercises, and the extant archive of Ulm models attests to the wide range of topological investigation. On the other hand, as we shall see, the network representation of building function became a characteristic Ulm exercise as well, a way to map the spatial operation of architecture. Indeed, Maldonado saw the fusion of technique and beauty as the heart of a reformed regime of design training, recalling Max Bill's admonition that "For us it has become self-evident that beauty cannot be developed from function alone. But we can demand beauty that is equal in quality to function."[53]

7.18   Lattice-oriented shell surfaces (Wand aus gitterorientierten Elementen), 1963–1965. Walter
Zeischegg with collaborator Ciril Cesar (Entwicklungsgruppe Z / Development Group Zeischegg).
Photographer: Roland Fürst, 1965. © HfG-Archiv / Museum Ulm.

7.19 Lattice-oriented shell surfaces (Wand aus gitterorientierten Elementen), 1963–1965. Walter Zeischegg with collaborator Ciril Cesar (Entwicklungsgruppe Z / Development Group Zeischegg). Photographer: Roland Fürst, 1965. © HfG-Archiv / Museum Ulm.

Beyond this curricular focus, Ulm was a nexus and promoter of topological developments within European design. The *Ulm Journal*, the official record of the school, relayed the topological research of Zeischegg and the various Ulm courses as well as experiments of other designers outside the school. Lech Tomaszewski, a Polish designer, was particularly engaged in such investigations, and his work, published also in the journal *Situationist Times*, no. 5 (an issue devoted substantially to knots and topology),[54] found a receptive audience at Ulm. Ulm and its fellow travelers thus represent one of the most thorough early attempts to inject topological ideas directly into the critical and pedagogical conversation of design.

Mastering Form, Global and Local

By the late 1960s, what had begun as gamesmanship at the edge of perceptual and formal impossibility matured into ambitious attempts to confront the *constructability* of hyperbolic and topological forms. Could twisting, curved volumes evolve from fantastic figments to precisely modulated assemblies orchestrated in perfect unison? Could curvature be amplified from sculptural curiosity to architectural tactic? Musmeci's bridge was indeed an early proof in concrete, but its monolithic casting skirted the stubborn problems of assembly from disparate parts that are more typical of architecture. With no mechanism to correlate couture details with overall gestures, topology's most radical architectural possibilities remained unrealized.

What design still lacked was a theory of topology not only as a generative engine for form but as a metaorganization to correlate every aspect of building design. Aided by topological methods, the remit of architecture could extend beyond designing bespoke individual buildings to systematically diagramming to rules of spatial arrangement. American designer Ron Resch (1939–2009) envisioned topology not only as a formal playground but as a governing law of complex architectural assemblies. In his essay "The Topological Design of Sculptural and Architectural Systems," presented at the National Computer Conference in 1973, Resch argued for new parts-to-whole relationships underwritten by relational topology. His critical insight was that a global skeletal network could coordinate the relationships among architectural elements, like the strings of a fantastically complex marionette. With this topological skeleton as an organizing rubric, geometric forms could be automatically rationalized for optimal fabrication and assembly.[55] These skeletons defined fluid and elastic bodies that self-adapted their subcomponents in a continuous and cascading way.

Resch's aim was the total customizability of architecture. He was convinced that computer-aided systems would enable endlessly elastic direct-to-fabrication structures selected and configured by clients on demand. He called this concept "soft prototyping":

> My work in the area of structure design, having been coupled with the computer as a medium of design and production, seems to suggest its possibility as a communications medium.... I believe from present experience that it is possible to reintroduce a "soft prototype"; to pay attention to the subtle variations of user needs; to conceive of objects as a continuously varying class of solutions to a continuously varying set of needs; and to use these needs as input to a transformation upon this topologically conceived object class such that it determines a specific set of instructions that will work within the variations made possible by automatic machines and process.[56]

Resch's soft prototyping freed architecture to transcend geometric specificity. Yet to achieve that, architects must embrace the design not only of buildings but of systems and software. This software, in turn, would become a factory of continuously variable building components.[57] In a prescient anticipation of parametric design and mass customization, Resch suggests that topology is capable of establishing variational relationships of part to whole, allowing complex ledgers of unique pieces designed or optimized to work together in a continuous fabric. A mathematical skeleton with sinews of differential transformations orchestrated constellations of parts and welded them into an architecture. For Resch, topology parsed an architecture of warped geometries into endless adaptive protocols.

The project of formal topology in architecture rested on a relational calculus of formations and deformations. Through a powerful language of diagrammatic transformations, disparate geometries were subsumed in a common and continuous meta-system. Vast geometric variation could be organized according to deeper topological logics. On the one hand, the result was an inexhaustible and continuous combinatorics of shape, a verdant garden within Meinong's Jungle. On the other hand, this fecund variety was matched by an equally expansive exact language of similitude and difference. It was the mathematical analogue of gestalt: form on the edge of integration, illusion, and disillusion.

Dual Topologies II: Topologies of Function

If formal topology overcame apparent impossibilities with lithe new architectural shapes, the functional topology of networks brought a regulating eye to the elemental problems of operational organization. Functional topology diagrammed cities and buildings as reticulated webs of streets, circulation pathways, and technological networks. In fact, foundational problems of urban planning—the logics of growth, distribution of resources, arrangement of buildings, and even the partition of the city itself—were often, at their base, topological.

Like two organisms locked in territorial struggle, buildings and streets create reciprocal and opposing pressures that wrestle and distort the urban fabric. As street networks slice the city into cellular blocks, the demands of human occupation exert constant pressure to densify these cellular blocks to absolute limits. In extremis, these contending forces induce legible geometric regularities. Consider, for instance, the construction of Haussmann-era housing in Paris between 1855 and 1890. During this extensive renovation, housing projects were often developed on irregular city blocks bounded by violently cut boulevards. Unlike the characteristically piecemeal aggregations of *hôtels particuliers* which preceded them, these block-scale developments were conceived all at once. Developments that bordered Haussmann's pristine new boulevards were often irregular shapes and demanded standard yet geometrically compliant apartment floor plans. The higher the housing density of a site, and the more regular the cellular apartments within that floor plan matrix, the more mathematically inevitable the induced subdivision.

A particularly acute example is the work of architect-developer Paul Fouquiau, who built several dignified but extraordinarily dense housing projects for urban laborers. His project at the intersection of rue Eugène Sue and rue Simart, constructed between 1879 and 1882, was designed with a specific eye toward maximum density, subject to dimensional regulations of courtyards, light, and air access.[58] Though they folded topologically around the cornered eccentricities of the blocks, the individual units had remarkably standard plans. The unintentional but inevitable result of these strict rules is a building plan which is almost identical to the figure known as the Voronoi diagram, a precise mathematical method which calculates the regular subdivision of homogeneous cells within an irregular boundary. Though more common in tightly packed biological cells, this Voronoi diagram defined the liquid transformations of apartment plans. The geometric constraints latent in this process of urban subdivision were practically topological *avant la lettre*.

fig. 7.20

While designers such as Fouquiau intuited the cellular logics of network topology, they ventured no specific mathematical theories to justify their techniques. They had no need: they proceeded unselfconsciously from the obvious strictures of the geometric problem at hand. In negotiating the top-down constraints of block shape with the bottom-up demands of light and air, definite forms of geometry emerged deterministically. The inevitability of the result was a consequence of the intrinsic facts of the problem, not of a deliberate mathematical deduction. Yet the result was the same, and the quasi-mathematical results attest to the deep regularities which emerge when architectural and urban rules are taken to their logical limit.

The radical insight that mathematical topology might be at the root of urban morphology did not come until decades later, and then not from an architect but from a scientist. At almost the same moment that Max Bill imagined his endless ribbons, British crystallographer John Desmond Bernal (1901–1971), whom Maldonado had referenced earlier in his advocacy of topology at Ulm, argued that functional topology was the hidden morphological law of the city. While famed for his contributions to X-ray crystallography, Bernal had eclectic interests and vividly speculated on the relationship of science to culture. He was also fully conversant in the state of modern art and architecture, and in 1937 contributed an essay on the topic of "Art and the Scientist" to the anthology *Circle*, edited by the architect Leslie Martin, the painter Ben Nicholson, and the constructivist sculptor Naum Gabo.[59]

Bernal boldly speculated on the future of architecture in his influential essay "Architecture and Science" (1937) in the *Journal of the Royal Institute of British Architects*:

> The importance of topology is probably even greater for architecture than that of symmetry. . . . A problem of the greatest importance in any building is how people can get from one part to another, and this problem becomes of crucial importance in buildings of great size. . . . Practical topological analysis would enable the architect to choose a structure which achieved the greatest mutual accessibility compatible with factors of appearance, stability, and cost. . . . The main value of these mathematical disciplines in architecture would be to replace intuitive and haphazard solutions by others that could be rationally arrived at.[60]

What Bernal saw, even more clearly than many architects, was the intimate connection between topological connectivity and architectural type. The apodictic topological facts of circulation entailed equally definite facts about how specific types of buildings tend to be organized. Bernal would prove prescient and his thoughts would be influential in surprising ways for mid-twentieth-century architects.

By the 1950s, designers had assimilated enough ambient rhetoric from scientists like Bernal to argue for topological methods in an informed and technically specific way. In the spring 1955 issue of the *Student Publications of the School of Design of North Carolina State College*, the architect and engineer Robert Le Ricolais offered the modest text "Topology and Architecture" in which he extolled the analytic power of topological design. Taking Bernal's line of thought yet further, Le Ricolais's applications were specific: "In its relationship with architecture, at least for problems of a circulatory nature, topology plays not only a role on the economy of

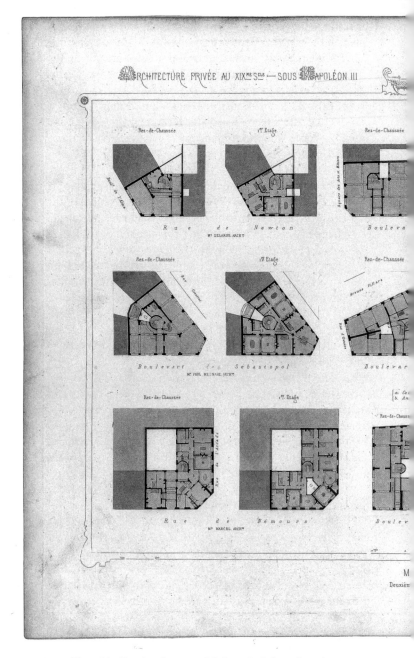

7.20   Proposal for unitized housing elements and their topological transformations.
Source: César Daly, *Maisons à loyer deuxième classe: parallèle de plans, n° 1, L'architecture privée au XIXe siècle, sous Napoléon III: nouvelles maisons de Paris et des environs* (Paris: A. Morel, 1864).
Bibliothèque nationale de France.

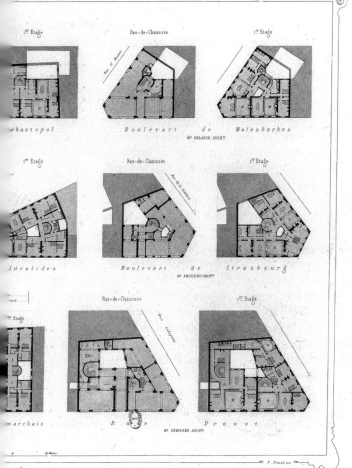

1ᵉʳ Étage · Rez-de-Chaussée · 1ᵉʳ Étage

Sebastopol · Boulevart de Malesherbes

Mr DELARUE, ARCHᵗᵉ

1ᵉʳ Étage · Rez-de-Chaussée · 1ᵉʳ Étage

Invalides · Boulevart de Strasbourg

Mr AMOUDRU, ARCHᵗᵉ

2ᵉ Étage · Rez-de-Chaussée · 1ᵉʳ Étage

marchais · Rue Drouot

Mr DEMIERRE, ARCHᵗᵉ

fig. 7.21

partitions, but a more far-reaching consequence is to be foreseen in the economy of displacements, i.e. time."[61] For Le Ricolais, topology could encode the temporal flow of building occupancy. The seed of topology as an ur-science of architectural organization began to take root.

Through the writings of scientists like Bernal and sympathetic engineers like Le Ricolais, designers saw a homology between the functional affinities of rooms and the spatiotemporal structure of architecture itself. The centerpiece of Bernal's proposed studies were weblike abstractions in which each point represented a room, and each line between points represented a permissible path of traversal within the building. With such diagrams, architects could reveal the geometric facts of inhabitation. A pioneering application of this functional topology was in the optimization of mission-critical spaces such as hospitals. In the United Kingdom, the establishment of the National Health Service in 1948 created an acute demand for objective studies on the performance of clinical environments. In 1949 the charitable Nuffield Foundation funded a series of studies to determine, quantitatively, the objectively correct arrangements for hospital floor plans. These studies approached architecture as a collaborative scientific enterprise: "A team of twelve, each a specialist in different subjects, under the chairmanship of an architect worked for five years on research into the design of hospitals."[62] That chairman was Richard Llewelyn Davies (1912–1981), a well-known architect trained at London's Architectural Association. No detail escaped Llewelyn Davies's scrutiny, from supply distribution logistics to the mechanical function of individual beds. But perhaps the most revelatory product of the 1955 study were network maps of hospital interiors that sketched the circulation relationships between the functional spaces and thereby reduced the floor plan to its most objective essentials.[63] Traced with a dense net of empirically documented paths, these studies unveiled the dendritic connectivity of inhabitation.

The Nuffield research revolutionized how architects tabulated the operation of space. No longer the subject of mere intuition, the use of buildings was diagramed by topological inscription. In the United States, a later 1960 study at Yale University used a similar method to observationally identify the

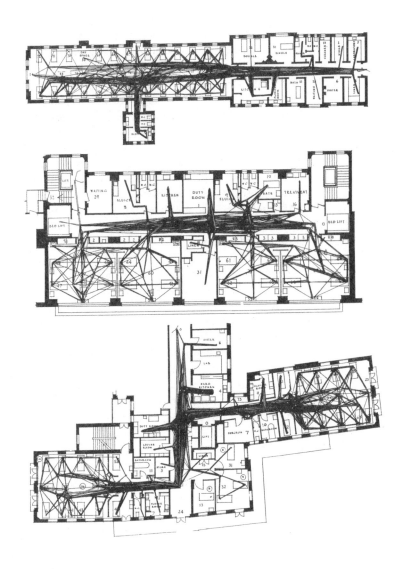

7.21   The patterns of nurse circulation at Westminster Hospital (bottom), National Hospital (center),
and Bradford Royal Infirmary (top).   Source: Nuffield Provincial Hospitals Trust, *Studies in the
Functions and Design of Hospitals* (London: Oxford University Press, 1955). © Nuffield Trust.

most frequent circulation paths of nurses within the wards of a hospital.[64] The proposed connectivity metric, known as the Yale Traffic Index, became a fundamental measure of the hospital type itself.

With projects like the Nuffield studies as a paradigm, the future of scientific design seemed to lie in the objective allocation of space under the governance of networks. The rigor of network topology was thus embraced in schools like Ulm HfG and Cambridge University at a moment when notions of design research and methods were ascendant. At Ulm HfG, for instance, as they were developing intricate topological surfaces, students also diagrammed the networked circulation of buildings on the HfG campus itself. Many of the key theorists who taught at Ulm, including the eminent operations researcher Horst Rittel, were convinced that topology was an essential basis for a truly modern mathematical theory of architecture.[65]

At Cambridge, the architects Lionel March and Philip Steadman applied graph networks to define topological invariants of floor plans, aiming at a Linnaean-like codification of building layouts. Both March and Steadman had undergraduate backgrounds in mathematics and approached the topic appropriately informed. They demonstrated the equivalence of floor plans through the corresponding equivalence of their underlying network structures, venturing an exhaustive graph-theoretic catalog of possible building floor plans.[66] This taxonomy was a combinatorial basis for objective plan comparison, one that was true to the topological tendency toward classification and was highly reminiscent of Tait's "orders of knottiness." March and Steadman also introduced a remarkably early instance of the so-called polygon skeleton, one of the invariant structures for shape discrimination that was legible in Fouquiau's early Haussmann plans. For March and Steadman, networks were not only a tactic for evaluating specific buildings but a strategy for comparatively ordering the whole of architecture.

fig. 7.22

As functional topology was encapsulated in digital techniques, it promised to automate an objective basis for architectural organization. Commercial ventures led the way, and during the 1970s, generative software tools were entrepreneurially developed under the auspices of several established architecture, interior, and furniture companies in the United States. A July 1971 issue

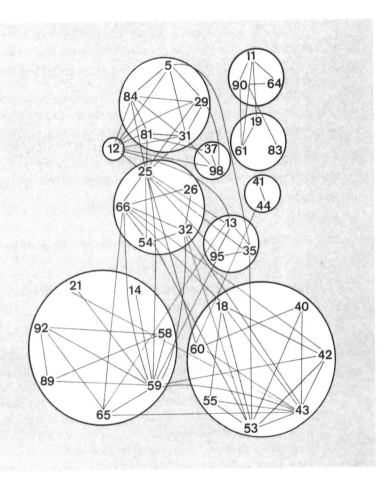

7.22 Diagrams of Herman Miller's computational system for workplace layout.
Source: "Designer's Utopia?," *Progressive Architecture* 52, no. 7 (July 1971): 87.
Courtesy of Herman Miller, Inc.

of *Progressive Architecture* devoted to the implications of digitization highlighted offices with bespoke software to generate floor plans by networks of spatial adjacencies. Even furniture designers entered the fray, as the workplace environments company Herman Miller developed an advanced digital organigram software they called the "Decision Resource Service" or DRS.[67] DRS took as input the organigram of a large organization—its departments, management, and report structure—and laid out that sociological diagram as a proxy for the topology of space itself. For Herman Miller, such tools totally reordered the human process of spatial planning: "The interior designer's role, once limited to working within the client's program, has now expanded to writing the program based on an evaluation of the client's needs."[68] A topological view of spatial and social organization thus began to rewire not only the products of design but the discipline of design itself.

Allometry and the Topology of Growth

The network topology studies of the Nuffield Foundation or Herman Miller peeled back the skin of architecture to reveal the skeletal form of circulation that shaped buildings and, by extension, cities. As Mark Wigley noted in his essay "Network Fever," by the early 1970s, "networks were now the beginning rather than the end point of city form."[69] But networks also provoked much larger questions of how spaces, buildings, and cities transmute themselves as they grew over time. The science of *allometry*, the study of this differential, network-constrained proliferation, addressed these questions mathematically while bringing architecture ever closer to allied sciences like biology. First popularized by biologists like D'Arcy Thompson (1860–1948) and Julian Huxley (1887–1975) in the early twentieth century, allometry originated as a tactic for analyzing organic growth. Allometry applied topology to identify invariants and transformation vectors through the growth process of organisms. Inherently transdisciplinary, allometry was revived in the mid-twentieth century to explain the processes of formal differentiation across fields from biology to architecture.

The common enigmas of systemic growth made scientists and designers comrades in arms, fellows in a search for growth's deep geometry. The self-declared Harvard Philomorphs were one such cross-disciplinary group, including biologists, metallurgists, and computer scientists, that coalesced around problems of topological growth. The informal cadre consisted of the eminent biologist Stephen Jay Gould (1941–2002), the design theoretician Arthur Loeb (1923–2002), the metallurgist Cyril Smith (1903–1993), Michael Woldenberg of Harvard's Laboratory for Computer Graphics, and the urbanist Ranko Bon (b. 1946), among others. Gould, the de facto leader of the group, was particularly taken with architecture as a case of evolutionary morphology. In his remarkable paper "The Spandrels of San Marco and the Panglossian Paradigm: A Critique of the Adaptationist Programme," he argues that apparently idiosyncratic architectural elements like the gothic spandrel are actually quasi-biological adaptations to eccentric topological conditions.[70] Architecture, topology, and evolutionary biology converged in a common morphological worldview.

The designers among the Philomorphs brought topological ideas to bear on theories of growth and form in architecture and urbanism. In 1972, Ranko Bon completed his thesis at Harvard's Graduate School of Design, under the advisement of Woldenberg and Gould, on the application of allometric principles to urban growth and the programmatic organization of architecture.[71] Bon used a statistical analysis of hundreds of geometric scenarios to derive what he called the "topologic structure" of urban forms.[72] In his striking argument, he empirically calculated several "nearly perfect" invariants which remained virtually constant as buildings or conurbations changed in scale.[73] The invariants analyzed spanning networks and trees of circulation identical to those drawn at Ulm and discussed by March and Steadman at Cambridge. For Bon, the same skeletal networks were deterministic vectors of urban growth and expansion.

If formal topology danced at the edge of rationality, the functional topology of networks embraced by designers like Bon derived architecture and urbanism from the irreducible facts of spatial arrangement. As designers searched for the invariant essence of architectural organization, networks held a promise of ground truth.

Paradoxes, Seen and Drawn

In Alexius Meinong's peculiar ontology of chimerical entities, impossible objects were not completely lost.[74] Meinong argued that they yet enjoyed a sort of "quasi-being," thanks to the odd disjunction between reference and object: the round square, he argued, was round whether it existed or not.[75] The bizarre gestalt puzzles at the limits of perception and mathematics inhabit something akin to this zone of quasi-being: the impossible objects to which enigmatic topological drawings referred were caught between dimensions, perceptions, and realities.

In time, the computer became a factory for impossible objects. Digitization allowed the sundry technical complexities of topological architecture scrutinized at Ulm, Cambridge, and Harvard to be stitched into the code of digital instruments. In this way the separate topological discourses of impossible objects and network representation were irrevocably interwoven. But digitization also raised strange new questions about both machine and human perception. Indeed, even for architects like March and Steadman, the classificational power of topology had its endgame in machine representation "where pattern recognition has become a central subject of research."[76] How would seeing machines interpret the cryptic topological objects which so confounded human intuition?

In the new world of computer graphics, impossible objects became gestalt tests of machine vision. In their 1958 paper "Impossible Objects," the father-son pair of psychiatrist Lionel Penrose and mathematician Roger Penrose defined impossible objects as drawings in which "the whole figure ... leads to the illusory effect of an impossible structure."[77] Computational researchers took the Penroses' impossible objects as Rorschach tests for the computer itself, or as a Turing test for the computer's spatial perception: measures of its ability to decode nonsense drawings.[78] The structure of such drawings, like that of Meinong's speculations, was linguistic: American mathematician D.A. Huffman (1925–1999), for example, offered "grammatical rules" for the machine interpretation of such forms in his "Impossible Objects as Nonsense Sentences."[79] Humans and machines were decoding, in parallel, the same language of topological ambiguity.

The long trajectory of design's topological experimentation had been all but forgotten by the 1990s, when topology was rediscovered as an engine for a new generation of formal and computational experimentalism. Designers reveled in a fluid architecture pulsing with new vitality. But at its root, this new fascination was, in fact, a recurrence of architectural obsessions originating in the 1930s. Then as now, the enduring promise of topology lay in its paradoxical capacity to be operationally precise but formally puzzling: impossibility and inevitability were two sides of the same coin, two views of the same enigma. Topology invoked the strange disorienting sensation of objects unbound from perceptual convention. Yet the science of topology was more than skin deep: beyond a superficial and esoteric exercise in styling, it seeped into the bones of architecture, furnishing a radical new substrate on which the entire relationship of parts to whole, individual to category, and invariant to transformation might be interpreted. The ultimate prize for the architect was to draw those paradoxes, capturing that moment of contradiction, building in the thick of Meinong's Jungle.

8　All You Need Is Cube:
The Political Economy of Grid Space

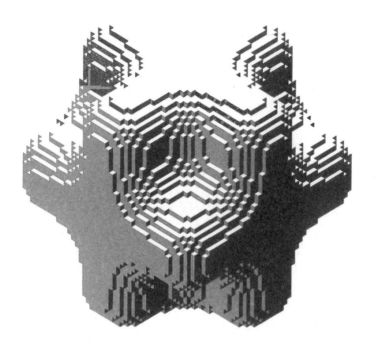

8.1   Voxelization developed by the Center for Cubic Construction.
Source: Centrum voor Cubische Constructies, *Cubic Constructions Compendium*
(Deventer, Netherlands: Octopus Foundation, 1970).

Buildings are economic projects as much as cultural ones, and design is inextricably bound to financial systems. As a framework for spatial organization, the cubic grid imported economic patterns of thought—the discretization of process, value, and space—into architecture. By partitioning space into a matrix of uniformly measurable elements, the grid carved space itself into a type of currency. Appearing at first in the 1930s as a passive playground for the so-called invisible hand, the cubic matrix in architecture evolved to become more like a referee: it regulated architecture as an active game of economic and even political dimensions. Through its evolution, culminating around 1970, the cubic cell iteratively and inexorably transformed from a passive unit of measurement, to a unit of simulation, and finally to a gamelike framework for economic or political choice. The cubic encoding of space exposed design to new metrics of comparison, fundamentally altering the resolution at which design choices could be made. Architecture was a chess board, a puzzle, a Rubik's cube.[1] Within the grid, gaming the system was not a rogue play but a design imperative.

The protean cubic matrix was an endless expanse of homogeneous containers for infinitely variable contents. Mathematicians recognized this endless flexibility as early as the 1890s when they first used the prima facie rigid cubic grid not as an end in itself but as a seed from which to generate undulating, twisting, and warped surfaces. German mathematician Hermann Amandus Schwarz's (1843–1921) method of extending minimal surfaces in cubic lattices, for instance, was a complete formal contrast to the Cartesian lineaments of the grid itself. Yet his drawings embedded complex minimal surfaces within indefinitely repeating cubic cells. For Schwartz, the grid was a necessary apparatus for the formal gymnastics of more complex surfaces. In an exactly analogous fashion, the grid, that simple homogeneous matrix, is a direct facilitator of the fantastically variable digital surfaces that have proliferated in architecture over the last two decades. It is an organizer of form but not a dictator of it, an endlessly adaptable scaffolding for almost any mathematical or visual language.

Beyond their geometric potencies, cubic matrices were also fences to bound and control risk, and thereby liberate creation. Every building project is a projection of the future and an actuarial

fig. 8.1

fig. 8.2

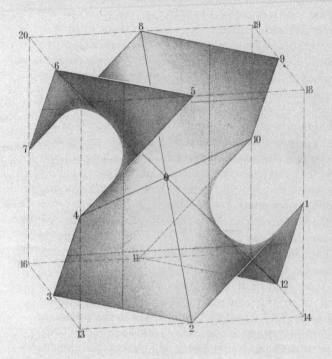

Verlag von Julius Springer in Berlin N.

Lith. Anst. v. C.L.Keller in Berlin S.

8.2   Minimal surface inscribed in a cubic armature.
Source: Hermann Amandus Schwarz, *Gesammelte mathematische Abhandlungen*
(Berlin: J. Springer, 1890), pl. 2, 4.

gamble on myriad factors outside the control of the architect. In his book *Economic Science Fictions*, William Davies pointed out that "Modernism, as a creative project, involved the construction of future worlds, cities, economic models and lifestyles, which would remove the technological and economic constraints of the present."[2] In containing risk, the cubic matrix could also unfetter the imagination—if the architect played by its rules.

Over the last century, the grid has acted as a tool of frequently contrary aspirations, perfectly adapted to advanced capitalism or open-source altruism, technocratic bureaucracy or social democracy. Like the contents of its cells, the associations of the grid were confoundingly heterogeneous. Even so, its political utility was irrefutable. At an urban scale, architect Pier Vittorio Aureli located the political potency of the grid in its capacity to "give permanent form to [the] possession of land and to define a template for coexistence."[3] Just as the grid subdivides space for private gain, it also provides the basis for shared politics. Equal parcels were the spatial manifestation of an aspirationally equal society. When the grid is spatialized and mapped into architecture as the cubic matrix, it becomes a plan for atomizing form into infinitely exchangeable and computable units. But could that atomization also bring freedom? The grid was a tacit social system, a mathematical equivalent of an idealized political economy.

Though the conceptual origins of cubic design run to the foundations of military strategy and the invention of wargames in the nineteenth century, the architectural implications of the cubic matrix only began to play out in the 1930s. Through the economic transformations of the following decades, cubic matrices persisted as an adaptable vessel of spatial partition. Architects enlisted it to discipline the building process and to leverage economic logics as catalysts of critical creativity. The cubic matrix was a design armature that underwrote diverse ambitions, an essential gambit to reformat architecture itself as an economic system.

Cube Roots

The financial crash of 1929 and the ensuing Great Depression prompted searching reflection on the economics of American life. Every aspect of the financial system seemed to rest on shifting ground. In exquisite relief, the realities of uncontrolled markets run amok showed yawning cracks in the quantitative systems used to manage and coordinate the American economy. The vast financial infrastructure of the building industry made it particularly vulnerable. Among architects, an urgent question was how the fragile practice of architecture could resist such forces. Yet, for architecture, there was reason for optimism. At the same moment, fresh modular approaches to construction cast the building as a controlled economy in microcosm. A modular building was not only a more rational design, but also evidence of a rationalized economics of building construction. If the building industry could be remade for the vast new demands for affordable housing, some designers hoped a more resilient and equitable society would follow.

fig. 8.3 Modularity was thus a bulwark against economic and social collapse, and the sugar-cube building was born of twinned desperation and optimism. It first appeared not in the drawings of an architect but in the theories of an American economist, Albert Farwell Bemis (1870–1936), who aimed to drag architecture into economic modernity for the good of society. Bemis drew on a broad interwar drive toward national efficiency in American manufacturing ignited in part by the 1921 report *Waste in Industry*.[4] Compiled by the Federated American Engineering Societies, the report assessed a range of domestic industries and took a particularly dim view of the nonstandard building process that was "tempered to suit the fancy of architects and owners."[5] It went on to categorize the wider systemic aspects of building waste, from inefficient financing to irregular labor practices.[6] Standardization, in particular the modular dimensions of construction components, seemed an obvious and necessary step to redress these deficiencies.[7] Industrialists like Henry Ford (1863–1947) and Frederick Taylor (1856–1915) had demonstrated a radical discretization of American labor and process. Bemis added to this a rigorous discretization of space.

Bemis understood that at the root of economic volatility was the precarious foundation of America's vast building industry, which was practically an economic system in its own right. He argued that the challenge of disciplining architecture for economic efficiency was less a matter of material optimization and

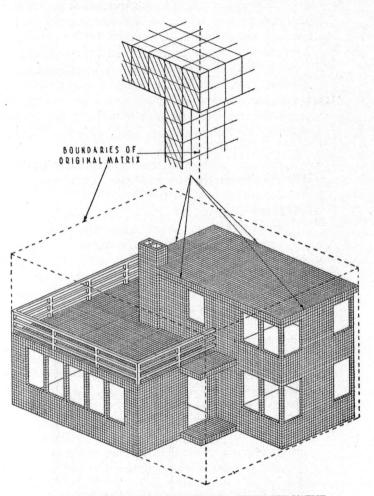

BOUNDARIES OF
ORIGINAL MATRIX

Fig. 19. THE HOUSE STRUCTURE DEFINED WITHIN THE MATRIX

8.3   A cubic discretization of a house based on Bemis's four-inch module.
Source: Albert Farwell Bemis, *The Evolving House*, vol. 3: *Rational Design*
(Cambridge, MA: Massachusetts Institute of Technology, 1934), 85, 154, 268.

more a consequence of organizational behavior. What architecture needed was a universal standard for measuring and designing space. Culminating with his three-volume work *The Evolving House*, Bemis proposed a radical modular system for domestic construction based on the uniform cube as the universal atom of architecture. Published serially between 1933 and 1936, *The Evolving House* encompassed an expansive scope from the ethnographically organized "History of the Home" in its first volume, an "Economics of Shelter" in its second, and finally a proposal for "Rational Design" of buildings in its third.

For Bemis, previous systems of modular construction, once full of revolutionary promise, had fallen woefully short:

> No one of them can be said to have offered anything like a complete scheme for rationalizing housing. They have aimed at mass production, to be sure, but in practice they have concerned themselves mostly with particular features of the house and with special designs of limited application. Their ideas are not thoroughgoing; they fail to effect the necessary drastic changes.[8]

To support this claim, Bemis's collaborator, engineer John Burchard, offered alphabetized criticisms of existing modular systems including those of modern architects like Buckminster Fuller, Richard Neutra, Frank Lloyd Wright, and Walter Gropius. Burchard argued dryly, for example, that Wright's famous concrete textile-block construction "does not seem to go very far toward pre-fabrication."[9] Burchard was particularly skeptical of the commercial impact of Fuller's famous Dymaxion house: "Dymaxion, in short, should be regarded not merely as a house but as an expression of an entirely different philosophy of living—as such, a corresponding amount of sales resistance must be admitted."[10] International systems from England, Germany, and France were also weighed and often found wanting.

Bemis answered these half measures with his own ambitious response: an exhaustive matrix of space-filling cubic partitions to order architecture. This cubic subdivision would regulate the granular dimensions of every constituent building element in the entire construction industry. Individual beams, columns, finishes, and components had to conform to this module in order to be interchanged seamlessly: "In other words, the parts of a house may be mass-produced and mass-assembled if they are designed on a cube module and therefore may be represented within the complete structure by multiples of this cube."[11] The cube became a ruler for both the individual building and the entire building industry.

fig. 8.4

The architecture that conformed to this cubic armature could be infinitely permutable, even fluid: "If the house can be designed on the basis of a cubical module, with its identical linear dimension in all three cross-sections, an identical variation of all members in all three dimensions will be possible and, consequently, the absolute interchange of similar parts and their standardized assembly in all three planes."[12] The cubic matrix ordered the ideal bounds of individual components, as well as the dimensions of detail connections and joints, all subject to a standard rubric. But this ideal system had to become physical, to be "translated into terms of existing materials." Bemis acknowledged that because "an unvarying homogeneous substance requiring no tolerances for expansion and contraction is not available, the problem of dimensional variation must be met."[13] Joints raised particular challenges, and Bemis attended to their resolution fastidiously, sometimes employing recursively subdivided secondary jointing cubes for connections.[14] He even qualified distinct cubic modules for distinct construction types: four inches per side for timber, eight inches per side for masonry, and so forth, all deduced from the specific minimum resolution of construction suited to each material. These abstract gridded scaffoldings were a means not only to measure individual components but to induce a parts inventory in harmony with an overarching production processes.

The cubic modular system was both a way to organize material and a strategy for standardizing and regulating information, particularly the numeric information of dimensions governing building components. Implicit in Bemis's system was the need for a sophisticated orchestration of modular arithmetic. Individually, the cube provided "a unit of measure, an actual quantity as simple, self-evident, and elementary as arithmetic."[15] Yet in aggregate, coded as differentiated functions and individuated subcomponents, only complex arithmetic operations could align and

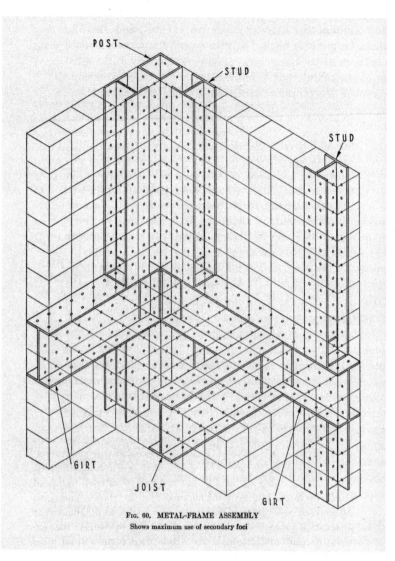

FIG. 60. METAL–FRAME ASSEMBLY
Shows maximum use of secondary foci

8.4 An example of metal framing construction resolved according to a cubic module.
Source: Albert Farwell Bemis, *The Evolving House*, vol. 3: *Rational Design*
(Cambridge, MA: Massachusetts Institute of Technology, 1934), 154.

coordinate cubic assemblies. An extensive machinery of simultaneous equations was required to tame such an intricate organism.

The grid's regulation of material implied a wider regulation of political and social systems. Bemis argued that only a system that touched every aspect of building—from design to finance—could make architecture conform more harmoniously to the economic processes with which it interacted:

> Its objective is to provide suitable homes at reasonable cost for the entire population.... In the process of attaining this objective, finance, employment, purchasing power, rent, marketability, architecture, building regulations, the standard of living that is, all the additional factors involved in the provision of housing, with the possible exception of land and taxes will become rationalized.[16]

Every social dimension was tacitly implicated in Bemis's modest cubic grid, which he believed would "have an effect on our whole economic structure."[17] It built a common armature for the disparate physical and social forces shaping buildings, and thereby articulated spatial rules of industry in microcosm. Bemis's cube was the building block for a whole political economy.

The rigor of Bemis's work proved enormously persuasive and accelerated both national and international efforts toward dimensional standardization. In 1939, the American Standards Association (the predecessor of today's ANSI, a body that sets voluntary national standards) initiated a project to develop and promote a modular standard based on Bemis's work.[18] In 1953, the US government began promoting Bemis's vaunted four-inch module through publications of its Housing and Home Finance Agency.[19] For a time, it seemed the cube would be enshrined as a geometric basis for the entire American building industry. Architects embraced the simplicity of the cubic standard, and it became de rigueur in the design of manufactured housing projects of the time. It was alluring enough that Konrad Wachsmann and Walter Gropius's Packaged House System of 1942 was organized on a "minimum cube" module of four feet, which was almost ceremonially inscribed in the drawings produced by their General Panel Corporation.[20] By 1954, the momentum around modular building had become truly international, with collaborative reports by ten European countries and the United States declaring the need for systems along Bemis's model.[21] Ultimately, cubic armatures were cast as an inevitable response to the unpredictable contingencies impacting housing not just in the United States but universally. Economy and geometry were correlated on the common playing field of the cubic matrix.

The Cubic Chemistry of the Welfare State

Across the Atlantic in Britain, the cubic module was used not only to regulate building dimensions but also to simulate the spatial futures of governments and nations. In Britain, the postwar rise of the welfare state encouraged technocratic economic policies and rigorous mathematical planning. This reciprocity of economy and government was fully embraced by the British architect Leslie Martin (1908–1999), one of the most influential British design intellectuals of the 1950s and 1960s.[22] Where Bemis saw the cube as a universal ruler for space, Martin saw cubic architecture as a proxy to simulate complex spatial and social scenarios. In a series of commissions and in his research at Cambridge University during the 1960s, Martin deployed the cube as an experimental framework for the behavior of space itself, using it to schematically sketch, permute, probe, and optimize vast numbers of spatial options for public works projects. Cubic frameworks interrogated the living pulse of space as architecture's natural lab instruments.

figs. 8.5, 6 Martin's interest in the cube as a generative tool was actually piqued from the foundational role of crystal lattices and cubic matrices in physical chemistry. While reviewing John Stanley Durrant Bacon's *The Chemistry of Life* (1944), Martin's student and collaborator Lionel March (1934–2018) discovered by happenstance a diagram which illustrated chemical recombination as a checkered grid.[23] Historians Adam Sharr and Stephen Thornton observed that these grid diagrams resembled combinatorial comparisons of volumetric envelopes or perhaps real estate propositions of gridded developments.[24] In the diagrams, Martin and March sensed potential for exhaustive dimensional permutations as objective ways to explore and evaluate design. Martin noticed a direct correlation with the discipline of econometrics,

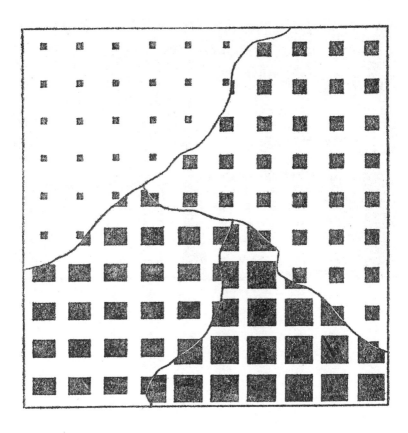

8.5 Diagram illustrating the proportionality of nutrient density in various foods, which partly inspired Lionel March and Leslie Martin's cubic permutations.   Source: John Stanley Durrant Bacon, *The Chemistry of Life* (London: Watts Thinkers Library, 1944), 9.

8.6   One of Martin's combinatorial exercises on solar access.
Source: Leslie Martin, Colin Buchanan, and Ministry of Public Building Works,
Whitehall: A Plan for the National and Government Centre (London: H.M. Stationery Office, 1965).
Open Government License, UK National Archives.

what he called "the 'mathematical movement' in economics," and its "structuring theory around which growth and change can develop."[25] By generating multiple options of subtly changed configurations of cubic cells, architects could iteratively test and evaluate design hypotheses experimentally.[26] What were the most efficient volume-to-area ratios for the building envelopes? What was a configuration of volumes that minimized shadow? Surprisingly, Martin found his analyses particularly salutary when considering the human scale: the experiences of light, of shadow, of temperature, or of sound in a given spatial arrangement were geometrically determined and could thus be quantitatively analyzed through cubic permutations. Scalar transformations of cubic grids, when understood through the embodied experience of individual humans, were not abstractly neutral but had marked phenomenological qualities. Cold cubic matrices thus ironically became a heuristic for warm-blooded embodied experience, teasing out what Martin hoped would be a humane, equitable, and objective means to plan the public spaces of government.

For Martin, the combinatorial machinery of the grid was merely an exact way to make manifest and comparable a gamut of alternative futures. "The ultimate problem for the profession," he argued, "is that of setting out the possibilities and choices in building an environment."[27] Martin elaborated his plan for an analytic architecture in his influential essay "The Grid as Generator." Originally delivered as a lecture at the Harvard University Graduate School of Design in 1966, this text set up architectural and urban problems as puzzles to be cracked through measured proportional variation of cubic clusters. For Martin, the pervasive presence of the grid in urban planning "sets out the rules of the environmental game," and he came ready to play.[28] That game was to define the parameters of families of radically different distributions of space and land in terms which were objectively comparable. Martin was searching for a practice of spatial analysis that "always rests in a simple demonstration of a mathematical fact."[29] Martin's space planning was thus a hybrid between a territorial game and a bureaucracy of space, sponsored by the geometry of the grid.

287

The World, Cubed

By the 1960s, the advent of standardized shipping containers began to radically simplify and cheapen all modes of commercial freight, fueling a relentless flow of dimensionally regularized goods.[30] Leslie Martin's top-down vision of the cubic matrix as a singular spatial analyzer was already at variance with the increasingly bottom-up economic reality of market-driven containerization, commerce, and transport. Uniform shipping containers had their roots in the transport of mined ores in the nineteenth century, and in the twentieth century the same basic ethos was applied to every conceivable product.[31] National and regional standards in the United States and Europe that had developed from the 1930s to 1950s converged toward global regulations for intermodal shipping containers promulgated by the International Organization for Standardization (ISO) in 1967.[32] Distances collapsed and containerization erased the friction of global capitalism to hasten a postwar consumption boom.

fig. 8.7 For architecture, containerization implied that every bounded volume of space could be assigned a price: it was a fungible asset like any other. If architecture could be deconstructed efficiently and rationally and into containable units, it could be packaged and shipped. In the new economy of limitless convertibility, even the once-static elements of architecture could be circulated and exchanged. The logic of recursive subdivision could be repeated ad infinitum as a multiscalar nesting of fractal cubic tapestries.

When Bemis and Martin each envisioned the cube as metric or analytic unit, they never anticipated it as a literal tectonic element. They would have seen the misinterpretation of this abstraction as a gross category error. Yet containerization seemed to open exactly that possibility to young designers. In the late 1960s, the Dutch design collective Centrum voor Cubische Constructies (Center for Cubic Construction, or CCC) embraced the geometry of extreme containerization as an obsessive design framework verging on a total cosmology. Furniture, typography, self-portraiture—all of visual culture was raw material for cubic reconstitution. The CCC's two founders, William Graatsma and Jan Slothouber, had worked since 1955 for the Dutch State Mines

8.7   Examples of the hundreds of cubic objects and details developed by the Centrum voor Cubische Constructies.   Source: Centrum voor Cubische Constructies, *Cubic Constructions Compendium* (Deventer, Netherlands: Octopus Foundation, 1970).

developing exhibitions with design elements that evoked the containers used for rail transport of extracted ores. It was an unusual career path to be sure, but one which opened their eyes to the capacity of the cubic container as an architectural element. Instead of seeing the cube as an abstract economic volume, Graatsma and Slothouber argued that the cube itself could be the literal building block of architecture. As they described their undertaking:

> CCC is the incorporation of our common strivings toward an ideal purpose in the field of design. CCC is not a commercial undertaking. CCC's objective is to develop and spread definite ideas about the application of cubic regularities. ... CCC has in view the promotion of form experience and form insight, the rationalization of design methodology, the promotion of economic industrial fabrication of objects for use and the promotion of the user's creativity.[33]

Graatsma and Slothouber affirmed the cubic matrix as a tool to both generate and grade architecture. It was "a rational approach to design problems, by which motives and results can be argued and judged more exactly."[34] The grid was a means to bend economic logic toward free creative production with precision.

More than anyone before, the CCC made the lone cube an entire universe of design exploration. They ignored its stubbornly mute impenetrability, instead unwrapping and cracking open the cube, concocting endless alternatives for its contents, structure, and detailing. They also turned the cube's rationality on its head, taking its boundless combinability as a wellspring of creative invention. Like Bemis, the CCC's starting point was a chosen invariant dimensional module: "All sizes used are based on a cubic system in which a module of 700 mm is continuously halved or doubled. ... All constructions used are based on a cubic system in which a maximum of identity between the composing parts has been realized." Yet the CCC did not confine themselves to that dimension. Recursive nesting, fractal subdivision, and centrifugal aggregation created an endless gradient of cubic frames at every scale, from furniture to the city. Some constructions were even abstractly scaleless, such as their intriguing cubic discretizations of various geometric forms that anticipated the low-res digitizations of the near future.

The magnum opus of the CCC's cube fever was a combinatorial sourcebook for cubic architecture, the perfectly square 428-page *Cubic Constructions Compendium*, which they described as "a program of possibilities" in cubic form.[35] It was a manifesto for open-source geometry as much as it was a radical democratization of architecture. Published in 1970 for the occasion of the CCC's exhibition at the Venice Biennale, the *Compendium* was a vast and rigorous catalog of hundreds of articulations of cubic form. It might even be called a geometric cookbook: it was an index of recipes for cubic application, intended for utility, pleasure, and improvisational remixing. Graatsma and Slothouber saw these expansive combinations as a work of communal ownership, an open-source geometry codebase *avant la lettre*: "CCC has put aside all claims of artistic ownership and actively seeks to encourage the use of its ideas by everyone. This is perhaps the first time this has been done. . . . This activity has connections with ideals concerning the role of art and design in society, carried out before by the Dutch movement 'de Stijl' (maybe in a different manner)."[36] The invocation of De Stijl recalls Theo van Doesburg's yearning for a nonstylistic style, a design of universal and inevitable utility to all of society. The CCC inverted the economic logic of the grid and reoriented it as a political force for magnanimous communalism. It ascribed to the cubic grid an archetypal spatial equity. The cube matrix was more than a metric adjudicator. Promising a free open-source design frontier for everyone, it was perhaps the closest architecture could get to absolute social equity.

For all of its prolific production, at its core the CCC was a bold thought experiment rather than a pragmatic practice. In playing through their cubic game in all of its myriad scenarios, their work hovered somewhere between the idealistic and the absurd. While Bemis, the pragmatist's theorist, imposed a cubic ordering on the mundane universe of existing commercial building products, the CCC created its own new universe of architecture, anticommercial and open source, in defiance of that existing economic universe. It was a humanist affirmation of creative authorship both against and through the grid. Driven by an unwavering faith that the simplicity of the cube implied a universal social relevance, they cast the grid as the foundation of a new social utopia.

Computing the Cube

Space that was conceived as a gridded matrix was discrete, countable, and thus naturally computable. Logically, a cubic matrix was simply a list in which each cube was a binary bit, with contents that were full or empty, true or false, one or zero. The same qualities that suited the cubic matrix to economic applications also recommended it ideally as a framework for the digital encoding of buildings. By 1970, the eclectic cadre of designers, mathematicians, and planners that Leslie Martin had gathered at Cambridge University began to conceive the digital encoding of architecture through the cubic and combinatorial frame he had pioneered. Among the most energetic of these researchers was Lionel March, Martin's close colleague, who expanded the arithmetic of the cube into an expansive calculus of design representation. While March's wider computational contributions have been traced by others, what concerns us here are the specific ways in which March encoded and manipulated the cubic grid and its associated techniques as a universal arithmetic of architecture.[37] March's hybrid training included studies of both mathematics and architecture as an undergraduate at Cambridge, graduating in 1959. He was a fellow at the Joint Center for Housing Studies at Harvard and MIT between 1962 and 1964, and worked with Leslie Martin on his pivotal proposal to raze and rebuild Whitehall, the London administrative center of the British government bureaucracy. The germinal research of that project and its iterative testing of cubic scenario models became a seed of March's calculation of form and encoding of space.

fig. 8.8

March was literate in and drew on historical work surrounding cubic form, particularly Bemis's theories and the syntax of aggregated cubes promulgated by the crystallographer René-Just Haüy. What was yet missing, and what March and his colleague Philip Steadman hoped to lucidly furnish, was an algebra of architectural manipulation that could comprehensively orchestrate the allocation of an arbitrary space of cubes. They fused and extended their predecessors' ideas by pairing them with advanced

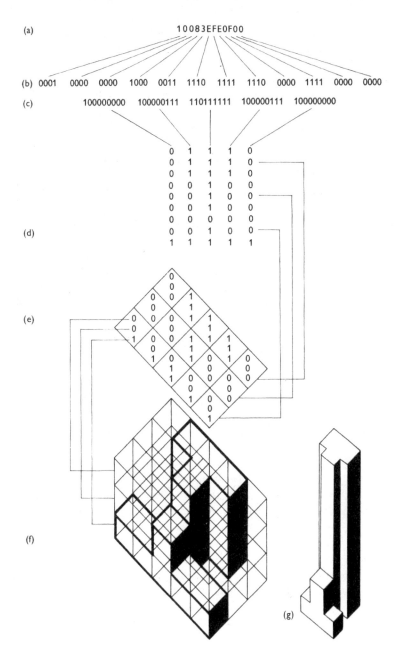

8.8 Lionel March's binary encoding of cubic matrix building forms.
Source: Lionel March, "A Boolean Description of a Class of Built Forms," in *The Architecture of Form*, ed. Lionel March (Cambridge: Cambridge University Press, 1976), 63.
© Cambridge University Press 1976.

techniques of discrete mathematics and modular arithmetic.[38] In particular, they cleverly applied Diophantine equations to choreograph the myriad wall positions and thicknesses in the cubic module structures of buildings, literally generating architecture from formulas.[39] Diophantine equations are systems of multiple linear equations with integer coefficients that are solved simultaneously as a system. Since their solutions were integers instead of continuous real numbers, Diophantine equations controlled the dimensions of architecture in a stepwise and discrete way, ideal for modular or cubic construction. Such equations were calibrated to produce a unitized solution fitting the variables at play. March and Steadman adapted them to solve the burdensome arithmetic of optimizing building components: "In the architectural context, we can imagine a variety of situations in which linear Diophantine equations might arise. What sizes should prefabricated wall panels be produced in so as to allow the greatest flexibility in design? In an office block ... what combinations of widths of offices can be fitted along the length of the building?"[40] Diophantine equations became a skeleton key to unlock the strict geometries of modular grids. In trained hands like March and Steadman's, simultaneous equations became a coded game, a new modular arithmetic of plan making.

Beyond its use as a lattice organization for space, the cube was an ideal atom for the digital encoding of buildings. The binary character of the checkerboard cubic framework transmuted any building into a sequence of zeroes and ones, the native electronic language of the computer. March and Steadman mapped this low-resolution partition into a unique hexadecimal code. In effect, they divined a method for encrypting digital facsimiles of buildings in strings of alphanumeric characters. They illustrated the method specifically for Ludwig Mies van der Rohe's Seagram Building, among others, showing that entire architectures could be rewritten in ciphers of machine-readable character strings. Where the CCC plumbed the tectonics of the cubic detail, March and Steadman theorized the logical behavior of cubic space in totality. The arithmetic they established at Cambridge laid out how buildings could be geometrized, digitized, and calculated.

Consciously or not, March and Steadman's selection of the Seagram building as an object of cubic discretization thrust Mies's methods full circle into a wider debate about architecture as economic commodity. Seagram's gridded curtain wall brought this conversation to a fine point. Reinhold Martin identifies the use of the modular curtain wall as an architectural tactic to negotiate mass industry, commodity, and seriality.[41] On the one hand, he notes the contemporary "physiognomic cataloguing of metal and glass curtain wall types," interchangeable systems of discrete industrial commodities reminiscent of Bemis's dream of standardization.[42] On the other hand, he notes that the undifferentiated framework of the curtain wall became a foil and container for the extreme heterogeneity of activities contained within it. Martin sensed the paradox of identicality and variability in the gridded frame, noting that it exists between "abstract autonomy" and "figurative plurality," between pure quantification and the contingent expressions of social life within it.[43] The cube was a black box. Therein lies the strange paradox of the grid: though perfectly uniform as a system, its contents are infinitely variable. The Seagram building, and the grid more generally, thus embodied the imbricated and fraught dialog between regularity and difference in mid-twentieth-century architecture and broader economic systems.

By the mid-1970s, at the time of March's encodings, the humble cube gained its own computational term of art: the voxel, a portmanteau of "volume" and "element," the three-dimensional analogue of the pixel.[44] "Element" evokes an atomic, almost chemical simplicity but also the possibility of molecular recombination. As the three-dimensional version of the two-dimensional pixel ("picture element"), the voxel became an atom of digital design and a quantum of total spatial inventories.

Design by Search

In their earliest guise, cubic encodings were an arcane bureaucracy of space, a technocratic invisible hand. The systems of Bemis, Martin, and March were intended less as tools for individual architects to design particular buildings and more as metasystems to regulate architecture itself. While their aims were socially progressive, the results veered toward centralized, hegemonic control. In countertendency, the CCC proposed a more egalitarian

fig. 8.9

perspective: the cube was a window into a society with free and equitable access to architectural knowledge. Yet liberated knowledge was not enough for a new type of architecture. To unleash the truly democratic power of a combinatorial architecture, the old model of the architect had to be overthrown.

The polymathic architect Yona Friedman saw the future of design and information as so mutually entwined that he believed architecture itself was poised to transform into an information-processing discipline, a cybernetic feedback loop between occupant and building.[45] With space sliced into digital cubic cells, the architect takes on the role of a game master who specifies the interchangeable data content of those cells. For Friedman, the feedback loop between space and content formed a process that cast the architect as an information intermediary: "This process was a simple one, made up of a transmitting station (the future user), a channel (the architect and builder together), the receiving station (the hardware, or finished building), and information return or feedback (the usefulness of the product made available to the client)."[46]

To accomplish this data-intensive new role, the architect of the future needed a new technology of information processing. Friedman's Flatwriter was that technology. The Flatwriter was a speculative computer for combinatorially generating and evaluating millions of distinct spatial plans for a particular dwelling, building, or city context. Consisting of a specialized fifty-three-key symbolically coded keyboard and associated display, it allowed users to directly input their preferences by selection rather than by consultation with an architect. The keyboard had a single key for each of several room shapes, door positions, and furnishing selections.[47] The user could simply type out the spaces of his ideal apartment. According to Friedman,

> Each future inhabitant of a city can imprint his personal preferences with respect to his apartment (flat) to be, using symbols which put in visual form the different elements of his decision so that the builder as well as his neighbors can understand what his choice is. In other words, this machine contains a repertoire of several million possible plans for apartments, knows how to work out instructions about the

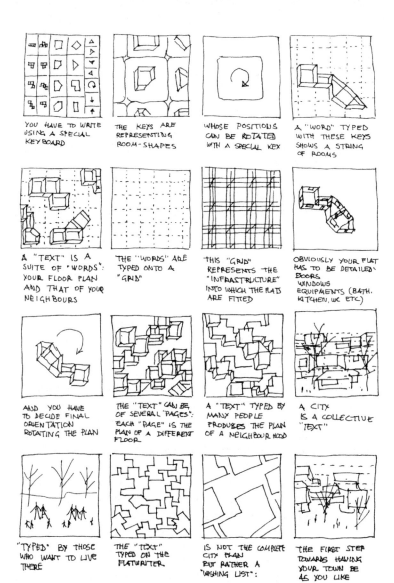

YOU HAVE TO WRITE USING A SPECIAL KEYBOARD

THE KEYS ARE REPRESENTING ROOM-SHAPES

WHOSE POSITIONS CAN BE ROTATED WITH A SPECIAL KEY

A "WORD" TYPED WITH THESE KEYS SHOWS A STRING OF ROOMS

A "TEXT" IS A SUITE OF "WORDS": YOUR FLOOR PLAN AND THAT OF YOUR NEIGHBOURS

THE "WORDS" ARE TYPED ONTO A "GRID"

THIS "GRID" REPRESENTS THE "INFRASTRUCTURE" INTO WHICH THE FLATS ARE FITTED

OBVIOUSLY YOUR FLAT HAS TO BE DETAILED: DOORS WINDOWS EQUIPMENTS (BATH. KITCHEN, WC ETC)

AND YOU HAVE TO DECIDE FINAL ORIENTATION ROTATING THE PLAN

THE "TEXT" CAN BE OF SEVERAL "PAGES": EACH "PAGE" IS THE PLAN OF A DIFFERENT FLOOR

A "TEXT" TYPED BY MANY PEOPLE PRODUCES THE PLAN OF A NEIGHBOUR HOOD

A CITY IS A COLLECTIVE "TEXT"

"TYPED" BY THOSE WHO WANT TO LIVE THERE

THE "TEXT" TYPED ON THE FLATWRITER

IS NOT THE COMPLETE CITY PLAN BUT RATHER A "WISHING LIST":

THE FIRST STEP TOWARDS HAVING YOUR TOWN BE AS YOU LIKE

8.9 Yona Friedman's operational description of his Flatwriter. "About the Flatwriter," 1967. Courtesy of the Fonds de Dotation Denise et Yona Friedman.

characteristic consequences of the way each future inhabitant would use an apartment, and finally, can determine whether or not the site chosen by a future inhabitant will risk upsetting the other inhabitants.[48]

The product of this interaction was not a plan per se but a "simple code that visualizes all elements involved in [the future inhabitant's] decision."[49] Instead of March's machine-readable alphanumeric cyphers, the Flatwriter translated architecture into a new visual representation that was readable by both humans and machines. The chess board on which this game of spatial moves and countermoves unfolded was a boundless voxelized lattice consisting of "a 3-dimensional modular grid structure, on several levels, raised on pilotis above ground level. This grid, while ensuring static stability, also contains the necessary mechanical and electrical services—water, electricity, gas, telephones, etc."[50] It was, in Friedman's words, "an infrastructure at the scale of the world."[51]

Friedman described the Flatwriter as a choice machine, an engine to tabulate recombinant inventories of preferences. In this machine, the architect was a facilitator of the encoding process and a check on the individual's most excessive antisocial tendencies. Friedman believed that "the task of the architect is to warn each user of the effect of each individual act of choice."[52] The Flatwriter threw into relief architecture as an orchestration of choices: choices of the client, choices of the architect, choices of the community, as well as metachoices about which choices were available or denied to each. It was a nearly perfect example of the concept of "choice architecture" from behavioral economics: the definition of the "context in which people make decisions."[53] For Friedman, the advent of interfaces for the scenario modeling of individual and community preferences elevated the building architect to the more expansive role of choice architect, marking maps of possible collective futures.[54]

The architect set rules not only of the building but of all the choices, of all participants, leading to the building. Friedman explained: "Once these rules are defined, we shall construct a combinatorial list of all the things that the rules of the mapping generate. And this list will correspond, element by element, to the repertoire which we wish to establish."[55] The product of his game was a vast cubic universe of floor plan possibilities, each one evaluated by special numeric matrices. Each matrix drew a differential map of the ease or difficulty of living in the associated floor plan. Friedman called these peculiar maps *isoefforts*. This neologism referred to topographic curves that mapped gradients of the "same effort" of uniform domestic work in a floor plan. In his parlance, isoefforts were graphic grids indexing the suitability of an architecture for a specific user. According to Friedman, the isoeffort was

> a sort of meteorological map that shows the fluctuations of the urban mechanism, and I think that it is as much a matter of public interest to everyone as is the weather report. All the applications cited here are based on this analogy with the weather map, from which the effort map differs only in that the city mechanism can be influenced by the acts and choices of the inhabitants while the weather, at least so far, is not subject to human will.[56]

Gradients and fuzzy thresholds of virtualized efficiency parameters, Friedman's isoefforts made the inhabitation of architecture a matter of simulated force dynamics.

The effort map was an economic optimization applied to interior architecture. It embraced domestic labor as the quantity to be tactically eradicated. A generalization of these maps, a kind of ur-map Friedman called the "Reference Matrix," could even be used to chart the space of architectural possibility itself: "This matrix is thus a simultaneous representation of all possible behaviors whether they exist or not. . . . This matrix reconstructs in a practical manner the field of reference of all possible histories of all possible organizations for cities."[57] Like the CCC, Friedman imagined the cube as the key to a total encyclopedia of space, an endless combinatorial library of architecture.

Friedman saw architecture in the information age as a communication system analogous to a telephone network. The Flatwriter provided the whole infrastructure in which the architect was a channel, the client a transmitter, and the architecture a receiver. He suggested that not only would the Flatwriter immediately enable the end user to design by selection, but that it would also allow frictionless communication with the building department for permit purposes and other construction automation tasks.[58] In this entire process, Friedman identified the architect as the fundamental bottleneck, since the architect's human capacities were simply not scalable to the speed and size of the client's demands or construction's complexities. Design through selection from vast libraries of ready-made architecture circumvented this limitation: "Instead of an architect, the future user encounters a repertoire of all possible arrangements that his way of life may require."[59] Friedman's machine supplanted the architect as epistemic liaison between client and building, while the architect is elevated to the role of editor of an extensive menu of spatial options. Design becomes less an act of divination on the part of the designer and more a systematic and ordered search by the client.

Friedman imagined that with the Flatwriter, architects would no longer work "for millionaires, but for millions of individuals."[60] The social function of architecture would be democratically liberated, with the grid as its revolutionary weapon. Friedman hoped to unveil the Flatwriter to the world at Expo '70 in Osaka, Japan, where visitors could test the future of design firsthand by printing their own customized apartment from a repertoire of three to five million flats. While the Flatwriter was never fully executed to this level, its conception contained the germ of a symbiosis between computer and human on the individual and societal scale. In effect, Freidman ventured a radical democratization of architectural combinatorics through the device of the cube.

All the World's a Grid

fig. 8.10

The Flatwriter turned architecture into a mix-and-match game of unitized tiles, albeit one with expansive social aims. But why stop at the door of the building, or indeed at the city? At a larger scale, macroeconomic systems might be mapped onto space as a sprawling discretization of terrain, flooding out in every direction over land, sea, and air. Just as the space of the building itself could be partitioned uniformly, landscape could be demarcated as groundwork for economic and logistic manipulation. Buckminster Fuller proposed exactly that with his World Game:

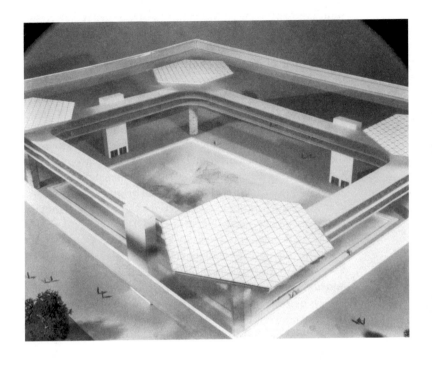

8.10   Fuller's proposal for a World Resources Simulation Center at Southern Illinois
University, 1966.   Source: Special Collections Research Center, Morris Library, Southern Illinois
University, Carbondale.

a systematic simulation not of mutual destruction but of mutual (and mutually interdependent) benefit.

In the 1960s, Fuller introduced the World Game to engineer large-scale multiplayer scenarios by which he hoped to reveal solutions to problems of systemic global interdependence. Part teach-in, part festival, part team sport, Fuller's "game that has no losers" co-opted the tools of economic strategy toward collective government at the scale of humanity. Fuller had spent time in the Navy and likely encountered strategic wargames through his training there. These wargames fixed a discretized grid over the territory of geographic contests, making Earth a zero-sum chessboard. Instead of using this game as an instrument for conflict, Fuller saw it as an instrument for cooperation. His approach recast the typical wargame: instead of the playing field encompassing a local territory like a battlefield, Fuller expanded the theater of action to the entire planet. His collated primer for a proposed World Design Science Decade served as a codex of planetary resources, an inventory of game pieces. The architect Roberto Bottazzi has called this inventory a definition of "the categories necessary to eventually compile the largest database possible on [the] world's industrialization."[61] It tabulated an enumeration of mined elements, a list of the atmosphere's chemical composition, surveys of usable water, censuses of human population, maps of electrical networks, and other material information to be used in the World Game. Fuller's inventories discretized resources into evenly measurable units as wargames discretized space into homogeneous cells. He believed this gamification was essential to match global resources to global challenges: "World gaming discovers the inventory of metaphysical capabilities that can amplify the life support effectiveness of the inventors of physical resources to accommodate all humanity."[62]

By collapsing economic, social, and ecological factors into a single grid, the World Game forced players to confront the unpredictable strategies of human actors as well as the vagaries of nature itself. The World Game was an implicit attempt to communicate and safely test responses to the erratic forces that had led to the 1929 financial crisis in the first place. Thus, the social and political dimensions of human relations were also woven into the grid of the game:

302

We can't see the atoms in motion; we can't see the stars move, though their motions are thousands fold faster than our fastest rockets; we can't see the trees grow; we can't see the hands of the clock move. Most important of all we cannot see the abstract weightless thoughts in the minds of other men. When we survey the total inventory of motions and informations which we can sense we find it to be very limited. The significance of all the foregoing is appreciated when we realize that it is only such phenomena as can be seen to be moving or changing by the public that are politically recognized and heeded. That is why public opinion and vote sampling has come into ever more reliable use.... We will soon learn popularly how to play the game to explore for ways in which we may use the world's resources so that we may be able to make our whole planet successfully enjoyable by all humanity without any human profiting at the expense of another.[63]

The definitive version of Fuller's World Game was a purpose-built stadium, decidedly modern and media-intensive, with projected maps of electrical networks, air transport, natural resources, agricultural bases, and population centers. Such a real-time economic and political simulation demanded an extensive electronic infrastructure. The mature version of the World Game was thoroughly computational and was intended to be moderated by the benevolent digital game master. Fuller explained: "We have organized at Southern Illinois University, and we hope it will soon be in operation, a $16 million dollar computer implemented program for playing just such a mutual success seeking game in a dramatically visible way. It will be so photogenic that it will become popularly and repeatedly broadcast on the world's TV circuits."[64] The World Game was the ultimate game show.

Fuller saw the World Game as a means of unveiling the hidden interactions of the planet's resources, politics, and economics for optimistic intervention. But he also saw it as defining a role for a new kind of computer intelligence, which would be a hybrid referee and boundless data catalog:

The human mind invented the computer as an extension of humanity's integral computer, information storing and retrieving system, the brain. The computer and the automated technologies it commands are about to take over all specialized tasks from humans, thus saving humanity from becoming extinct, for biological science and anthropology have learned incontrovertibly that extinction is always the consequence of overspecialization. Our World Game will be in effect a World Brain. It will free [the] world mind from occupations of brain slavery.[65]

Ultimately, Fuller imagined a World Game computer as a referee for democracy itself, aided by a mysterious bioelectrical telepathy: "world democracy is to be incorruptibly accommodated by continual electronic referendum, being progressively fed by subconsciously telepathic ultra-ultra-high frequency electromagnetic wave propagation, signaling subconsciously reflexed feedback attitudes toward specific propositions."[66]

Both Friedman's Flatwriter and Fuller's World Game emerged from an ambition to transpose economic constraints into a behavioral design framework. Fuller and Friedman acted as choice architects, drawing the horizon of a world for participants. Like Friedman, Fuller ascribed overtly benevolent qualities to the choice framework itself. The Game was a fair and idealized judge, the kind of objective political and economic moderator which was an inescapable, if sometimes hidden, element of other gridded projects. The ultimate goal of this vision was to encapsulate the details of economic life as a faithful but usable simulation of our imminent planetary challenges, broadcast to the world. The World Game was not merely a global optimization, but rather an engine for generating a right, good, and just society.

Pandora's Voxel

Around 1970, the cubic matrix reached a moment of apogee both as a simple graphic leitmotif and as a totalizing organizing principle. The Center for Cubic Construction's publication of their *Compendium* was completed and Yona Friedman hoped to unveil his Flatwriter to the world at Expo '70. While these designers foregrounded the voxel as an explicit object of study, it had ambiently permeated the work of many others as a stylistic trope. Adolfo Natalini and Cristiano Toraldo di Francia's Italian design collective Superstudio embraced the voxel in their Quadernas furniture series and in their *Continuous Monument* megastructure. More cybernetically, Nicholas Negroponte's 1970 installation *Software* at New York's Jewish Museum made the mythical invisible hand a visible robotic arm that dynamically rearranging stacked cubes in an abstract urban form, drawing inspiration from Friedman's work. Like the voxel itself, examples could be multiplied endlessly.

Just as it crested, the wave of cubic regularity felt increasingly exhausted and passé, and by the early 1970s the sugar-cube flood began to recede from architectural fashion. But while it was no longer such an overt visual motif in design, it persisted as an analytic substrate and methodological engine. The cubic matrix had the chameleonic quality that it could contain virtually anything in its neutral cells. It could be resuscitated and adapted in the service of virtually any exact (or sometimes inexact) discretization of space. It became an endlessly repeating partition for infinitely variable material, information, and fragments of architectural interest.

As a means to discipline space, the voxel still holds enduring charms for architects. It is agreeably blank, a mirror and a Rorschach test. Yet, for all its apparently anodyne character, the voxel grid is hardly neutral. In laying out a matrix, one circumscribes a world, what belongs to it and what is alienated from it. In that sense, the edge of the matrix struck a hermetic boundary, a threshold between ordered and unordered space. Gridded worlds were islands sealed safely from their surroundings, or placed in knowing confrontation with them. Unlike the triangulated survey drawing, which enmeshed architecture in the specificity of landform and location, the voxelized matrix unmoored architecture and freed it to be its own self-sufficient universe. Its graduated modules, like the serial markings of a ruler, abolish any outside metric. The matrix itself became the measure, economy, and government of all things.

9 Geometry's Mass Media: Broadcasting Technique in Architecture's Hyperbolic Era

9.1 The warped roof of the United Church of Rowayton,
photographed by Pedro E. Guerrero, 1962. © The Estate of Pedro E. Guerrero.

fig. 9.1 The enigmatic grace of thin warping surfaces ignited the imagination of midcentury architects and the general public in the 1950s and 1960s. Streamlined symbols of the space age, this new geometry of hyperbolic and ruled-surface forms was hailed by *Progressive Architecture* as a "New Sensualism."[1] The appearance of warped geometries was a turning point for the communication of architecture both in mass media addressed to the general public and in the more disciplinary media addressed directly to architects. While they had emerged earlier in the 1930s as an esoteric topic of engineering experimentation carefully guarded by patent law, hyperbolic and ruled-surface architectures burst into the public consciousness by the late 1960s as the pervasive shape of the future. Popular and professional media celebrated ever more daring forms, while designers and engineers raised the curtain on their technical secrets to the ever-expanding audience of trade publications. The design of complex geometry inflected and redirected the architectural use of media from the circulation of architectural images—of buildings, projects, and their affiliated representations—toward the circulation of mathematical tools and generative processes. This wealth of geometric information was, in turn, a prelude to do-it-yourself manuals that some designers self-published as a kind of democratic folk literature of technique, the once-guarded secrets available to the yeoman architect. Beyond the arresting qualities of the shapes themselves, the history of hyperbolic geometry in architecture raises issues of public and expert knowledge, insider and outsider cultures, and the specific mechanisms that promote formal innovation in architecture.

fig. 9.2 As both photographed form and mathematical technique, hyperbolic geometry circulated in popular and architectural media in distinct but symbiotic ways. Beatriz Colomina has argued that "it is actually the emerging systems of communication that came to define twentieth-century culture—the mass media—that are the true site in which modern architecture is produced and which it directly engages."[2] Popular mass media publications such as *Time* and *Life* magazines presented warped geometry to a broad audience, aligning it with wider lifestyle innovations and offering it as an appealing foil to the repetitive modularity of corporate modernism. Newspapers in the United

Where Tomorrow Begins..

Lambert - St. Louis Air Terminal
Architects: Hellmuth, Yamasaki &
Leinweber

*In the beautiful functional buildings
now rising, tomorrow is already
here . . . and so are tomorrow's
floor problems.*

The past 50 years has seen unbelievable progress in the development of new forms, new materials, new harmony of design — and the unveiling of new concepts of floor treatment which makes the modern functional floor practicable.

Hillyard, celebrating its 50th Anniversary, pledges significant contributions in continuing development of safe, economical floor treatments with proper built-in light reflective beauty. Limitless research, farsighted management and a nationwide staff of experts in floor treatment will guide Hillyard's second fifty years to new heights of service. Working closely with architects, builders, administrators and custodians, we are confident we shall help achieve yet higher standards for functional floor use and beauty in buildings "Where Tomorrow Begins."

HILLYARD CHEMICAL COMPANY
ST. JOSEPH, MO.
Passaic, N. J.   San Jose, Calif.

BRANCHES AND WAREHOUSE STOCKS IN PRINCIPAL CITIES

9.2   A flooring advertisement featuring St. Louis's 1956 air terminal and its complexly curved vaults.   Source: *Progressive Architecture* 38, no. 1 (January 1957): 10.

States and Europe hailed strange new forms as "the shape of roofs to come."[3] At the same time, a new disciplinary mass media, including professional publications like *Architectural Record* and *Progressive Architecture*, was expanding to vast new audiences. *Progressive Architecture* inaugurated their now well-known annual awards for cutting-edge design in 1954, as the press celebrated experimental new construction. In parallel with postwar commercialism and the popular embrace of novel forms, product manufacturers used advertising imagery to herald the dramatic effects of new hyperbolic geometries. The message was that a brave and forward-looking future required not only bold architectural geometries but also the commercial products to build them. As the practice of marketing itself became more central to commercial culture, savvy advertisers promoted hyperbolic design languages, and by association their products, as visions of high-tech contemporaneity. The streamlined curves of airplanes and Airstreams made their way into architecture.

The alignment of mediagenic visual novelty, technical education, and commercial interest produced a virtuous cycle that propelled warped and hyperbolic forms to a pervasive expression of a specific midcentury moment. In addition to being theaters for architecture per se, these media were also drivers for the realignment and reorganization of technical epistemologies that underpinned modern formal invention. As the mass media of popular and trade magazines exploded, they opened new ways to diffuse training and education around the geometry and mathematics of architecture. The cultural and disciplinary trajectory of hyperbolic architecture and its associated mathematical representations confirm the indivisible correlation between the communication of technique and the advancement of form. Geometry and mathematical knowledge were recast as cultural and commercial commodities.

Seeds of a Warped Language

A synthesis par excellence of architectural imagination and engineering precision, ruled surfaces provisioned architects with a rigorous but pliant geometric vocabulary. Seeming to defy both gravity and the strictures of rectilinear building, ruled surfaces offered a fresh alternative to more rectilinear modes of design. Among the range of ruled-surface forms that found their way into architecture, perhaps the most ubiquitous were hyperbolic paraboloids, or hypars, doubly ruled surfaces formed by the sweeping motion of a single line through space along a pair of straight-line rails. A mainstay of rooflines of the 1950s and 1960s, hyperbolic paraboloids as well as conoids, hyperboloids of revolution, and more exotic ruled and doubly curved surfaces were adapted for the roofs and walls of chapels, houses, libraries, sports venues, auto showrooms, and innumerable other typologies. Warped surfaces wrapped mid-twentieth-century architecture in undulating new clothes.

Novel though they then seemed, hyperbolic forms entered the architectural lexicon long before the twentieth century. In fact, architects had been aware of hyperbolic geometries since at least the seventeenth century, even if they had not fully applied them to the design of buildings. The British polymathic architect, astronomer, and geometer Christopher Wren (1632–1723) was an early investigator of hyperbolic and anticlastic forms. Wren's remarkable diagrams of hyperboloids of revolution for the grinding of glass lenses attest to the awareness of these forms as early as the 1660s.[4] As we have seen, the formal possibilities of ruled surfaces and hyperbolic geometries were more fully articulated in the early work of the French engineer Gaspard Monge and his students in the late eighteenth and early nineteenth centuries. We have already observed that mathematical models of hyperbolic forms were fundamental to engineering and even architectural education around the turn of the last century. By the fin de siècle, ruled surfaces became rare but not unheard-of elements of the architectural lexicon. A 1909 Yale University engineering text, for example, noted that hyperbolic paraboloids and the closely related oblique conoids were used by the "architect or designer of masonry structures" in "stairways, conoids of round towers, [and] arches."[5] Certain Beaux-Arts elements such as vaults or oblique passages also relied on the regularity of ruled surfaces for their resolution.

The rigorous construction of hyperbolic geometry did not widely emerge from these architectural manipulations but instead from the pragmatic imperative for structures of maximum span

fig. 9.3

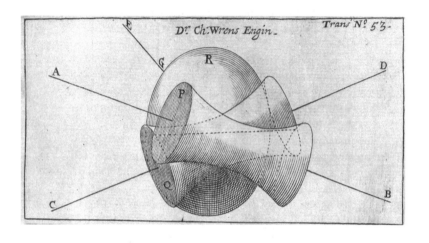

9.3   Christopher Wren's method for grinding lenses using hyperboloids of revolution,
presented to the Royal Society in 1669.   Source: Christopher Wren, "Engine Plus Figures I–VI,"
Philosophical Transactions of the Royal Society 4, no. 53 (1669).

and minimal material. Freestanding hyperbolic structures were thus first pursued systematically by engineers, who not only had the technical training necessary to calculate the loads of such spans, but also had likely been exposed to the formal vocabulary of hyperbolic mathematical models no longer in broad circulation among architects. The first tentative experiments emerged in light and diaphanous network structures, such as those developed by the Russian engineer Vladimir Shukhov (1853–1939). Graduating from the Imperial Moscow Technical School in 1876, Shukhov declined an assistant professorship with the eminent mathematician Pafnuty Chebyshev (1821–1894) in order to focus on engineering pursuits.[6] His professional career began in the then-booming rail and petroleum industries of the Russian Empire, where he honed a precise approach to the structural calculation of maximally efficient bridges and cylindrical storage tanks.[7] In addition to these industrial structures, Shukhov applied his geometric acumen to more architectural projects during the 1890s, notably the construction of an enormous tensile canopy and the first of his hyperboloid towers for the All-Russian Industrial and Art Exhibition of 1896 in Nizhny Novgorod.[8] Shukhov patented the underlying system and used it to design a series of revolutionary towers that translated the double ruling lines of hyperboloids of revolution into light compression members. Appearing as an almost direct transcription of Mongean principles or Théodore Olivier's mathematical string models, Shukhov's towers were elegant, almost unbelievably light presences. In the following two decades he designed dozens of towers scattered across Russia for water storage, radio transmission, and power line suspension, an early intimation of the structural potentials of hyperbolic form.[9]

Here a distinction must be made between hyperbolic geometry in the classical sense—ruled surfaces such as hyperbolic paraboloids and hyperboloids of revolution—and the axiom systems of non-Euclidean hyperbolic spaces which were emerging in the late nineteenth century. Non-Euclidean hyperbolic spaces were not discrete geometric shapes like spheres, cubes, hypars, or hyperboloids of revolution but rather were self-consistent axiom systems governing entire geometric logics. The Russian mathematician Nikolai Lobachevsky (1792–1856) famously demonstrated that his hyperbolic system of geometric axioms was just as consistent as the Euclidean system on which all of mathematics and the natural sciences had previously rested. An increasingly elastic definition of geometry itself—encompassing hyperdimensional forms and non-Euclidean axiom systems like Lobachevsky's—certainly contributed to a broadening of the vocabulary of rational forms in European mathematics and engineering in the late nineteenth century.

The architect and architectural historian Elizabeth Cooper English speculates that Shukhov may have been influenced by the non-Euclidean geometry proposed by Lobachevsky.[10] While this is difficult to definitively confirm or deny, it is also possible that more direct visual precedents for Shukhov's hyperboloid towers lay in Leonhard Euler's and Gaspard Monge's work on ruled surfaces. The theory of ruled surfaces was already in wide circulation in engineering schools throughout Europe, from Spain to Russia, in the same period. Russian engineering schools, in particular, were founded with the help of French engineers through an agreement between Tsar Alexander I and Napoleon.[11] Monge's descriptive geometry was integral to the curriculum, and a rigorous exposure to ruled surfaces would have furnished the necessary tools for Shukhov's inventions.[12] Regardless of Shukhov's source for his hyperbolic geometries, their realization undoubtedly required unusually refined mathematical and spatial intuition.

Hyperbolic geometry found its definitive architecturalization in the advent of gravity-defying solid thin structural shells of metal and concrete, born of painstaking experimentation by the generation of engineers that succeeded Shukhov. In the 1930s, the French engineer Bernard Laffaille (1900–1955) began to test the compressive capacity of monolithic hyperbolic shells. From a material and structural point of view, Laffaille found the anticlastic curvature of warped shells to be dramatically more efficient than typical post and beam construction. First in metal and then in concrete, he methodically calibrated geometric shape to maximize structural performance while minimizing material use. This experimentation led to the fabrication of conoid and hyperbolic surfaces in 1927 and 1933 respectively, arguably the first ruled thin shells to be truly structurally self-supporting.

For the earliest experimentalists, geometry was a legally defensible trade secret: Laffaille guarded his intellectual property with dozens of French patents.[13] Like Shukhov, he saw his remarkable innovations as a private asset to be strategically protected, a unique trade secret and a competitive advantage over other structural systems. The products could be publicized, but never the process. Laffaille had little reason to diffuse the rare and remarkable magic underpinning his constructions, which were developed with clear commercial motives.

Yet even at this early moment Laffaille was not alone. Hyperbolic geometry was already becoming a subject of intensely competitive international research and defensive patent registration. In Italy, Giorgio Baroni's 1937 Alfa Romeo factory in Milan showcased a doubly curved roof structure that rivaled Laffaille's.[14] Baroni was a consummate innovator, and he also developed a novel canopy structure which fused modules of hypars in a continuous, column-supported field.[15] He patented his work on "hyperbolic paraboloid or conoid form" not only in Italy but also internationally, including in the United Kingdom and the United States.[16] In Czechoslovakia, Konrad Hruban (1893–1977) also experimented with hyperbolic vaults, while in Spain, the engineer Eduardo Torroja's (1899–1961) work was arguably the most comprehensive early attempt to architecturalize these shells.[17] Torroja's Market Hall in the southern Spanish town of Algeciras (1934) and his better known Zarzuela Racetrack in Madrid (1937) established remarkable precedents for the use of hyperbolic geometry to dramatically roof public buildings.[18] Torroja also took pains to protect his work, filing at least eight patents between 1928 and 1940 on topics ranging from new types of pavement to surface structures of triangulated metal.[19] The collective work of these engineers validated the possibility of an evocative new geometry for architecture, but the knowledge associated with it was essentially private, produced from hard-won empirical tests and guarded by trade secrets and patent law.

A Hyperbolic Moment

In part because they were protected by proprietary patents, ruled surfaces remained the strict purview of engineering until the 1950s, when they began to be embraced and adapted as an alternative idiom of contemporary architecture. The expiration of the early hyperbolic patents coincided with a broader visibility and interest in architectural circles. This interest belied the fact that the surfaces themselves were naturally contrary to almost every received convention of architecture. Their continuous curvature prevented the flexible placement of doors or windows, and their smooth skins frustrated attempts to partition them into rectangular rooms. Yet the striking contours of hyperbolic surfaces were resonant with an age of technoscientific optimism. The space race was in full swing, and Albert Einstein's theory of general relativity recast spatial intuitions of the universe itself in terms of warped spaces. As the dynamic novelty of hyperbolic forms exerted a gravitational pull on mid-twentieth-century designers, architects of every sort—from lone experimentalists to established offices—were drawn to build with them.

Though they appeared nearly weightless, hyperbolic shells were intimately tethered to specific materials. The early research of engineers like Baroni and Torroja proved that hyperbolic structures were felicitously adaptable to concrete, brick, and wood, enabling structural solutions tailored to locally available resources. Methods of constructing concrete shells were particularly well suited to ruled-surface geometry, as ruling lines coincided with the planking of formwork. Engineers like the Spanish-Mexican designer Felix Candela took full advantage of the possibilities that ruled concrete surfaces afforded. During the 1950s, Candela dazzled the world with projects such as the Cosmic Rays Laboratory (1951) at the University of Mexico in Mexico City that showcased the power and versatility of structural shells by creating a vault with a material thickness of just over half an inch. Candela also collaborated with architects on hypar structures, such as his Capilla de Nuestra Señora de la Soledad (1956) with Enrique de la Mora y Palomar. He opened his technical secrets broadly to architects, contributing "Understanding the Hyperbolic Paraboloid" to the July 1958 issue of *Architectural Record*.[20] Masonry construction, with its long tradition of vaulting and close affinities to concrete, was also a natural fit for hyperbolic forms. In Uruguay, the intricate masonry shells by the engineer Eladio Dieste (1917–2000), notably in his 1960 Iglesia de Cristo Obrero y Nuestra Señora de Lourdes in Estación Atlántida, registered a new synthesis between ruled surface and discretized form. Built of thousands of meticulously placed individual bricks, these shells rendered modern forms in a decidedly traditional material.

Among the vanguard architects to embrace hyperbolic form in the United States, one of the most influential was Maciej Nowicki (1910–1950), a Russian-born Polish architect who acted as head of the design department at North Carolina State College in the late 1940s. Nowicki's most notable structure was the bold J. S. Dorton Arena (1951) in Raleigh, North Carolina, colloquially known as the Paraboleum. The roof of the Paraboleum hangs taut between two parabolic arch-beams which intersect at two points, one on either side of the building.[21] The light but capacious canopy was a revelation to architects across America and caused a sensation in the architecture press when it was completed in 1951. The Paraboleum was one of the key projects validating the new scale possible with new hyperbolic forms. Attracting many admirers and acolytes, it was probably the most influential early hyperbolic roof structure by an American architect.

figs. 9.4, 5

Nowicki died tragically in a plane crash in Egypt on August 31, 1950, before the Paraboleum was complete. In his stead, his protégée Eduardo Catalano (1917–2010) became head of the department at North Carolina State College.[22] Catalano had fully absorbed Nowicki's aesthetic sensibilities and technical knowledge pertaining to ruled surfaces, and was perfectly equipped to extend Nowicki's work in new directions. Long fascinated by the connections of mathematics and architecture, Catalano pioneered both the technical use of wood in hyperbolic geometry and novel forms of partnership with manufacturers and media outlets. Beginning in 1952, Catalano conducted a series of research studios, first at the School of Design at the University of North Carolina and later at MIT, which elaborated on the geometric abstraction and tessellation of hyperbolic structures into more articulated assemblages. He looked beyond the concrete shells of earlier engineers, toward wood structures and timber vaults. His first significant work, his own house in Raleigh, North Carolina, completed in 1954, was among the most visible homes of this period, photographed and referenced liberally in the professional

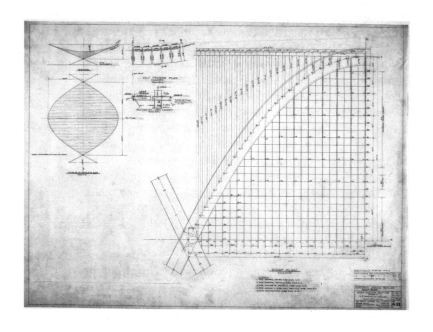

9.4   Roof plan drawing of the Dorton Arena (Raleigh, NC), showing one of the parabolic compression arches.   Source: Guy E. Crampton and William Henley Deitrick Papers and Drawings, 1928–1977 (MC00227), North Carolina State University Archives.

9.5   The Paraboleum shortly after construction. Division of Archives and History Photograph
Collection (State Archives of North Carolina Collection 4.1). Courtesy of the State Archives of
North Carolina.

and popular press. A minimal spanning surface of a single closed curve, it was the elegant essence of a new space-age dwelling.

Popular mass media feted the novel geometries of Catalano's hypar house as evidence of a visionary future. The house was covered rapturously in articles in *Time*, *Life*, and the *New York Times Magazine*, among others. A 1957 *Life* cover article announced "Tomorrow's Life Today," a radiant and imminent tech-enabled future: "the advances in technology have already begun to change the world—to put man's everyday living into a new realm of plastic houses and jet cars, to launch him vertically into the air on one-man platforms, to bring into his vision and hearing events as they happen anywhere on the earth's surface or deep under the sea."[23] Catalano's "Batwing House" was one of the "New Shapes of Shelter" emblematic of this bold tomorrow featured in the article.[24] Hyperbolic structures were among the "radical designs and structural concepts" that shaped a new popular consciousness of geometric design.[25]

Though Catalano's work was a media sensation, he retained an engineer's immersion in mathematical technique. His research embraced what he later called "inevitable form"—an architectural necessity born of mathematical deduction.[26] As precedents, Catalano specifically invoked the famous nineteenth-century Brill plaster models of mathematical surfaces, which he had been aware of since the 1950s, inverting and rotating them to suggest canopy, vaulting, and enclosure. Mathematical concepts such as minimal surfaces, hyperbolic structures, topology, mechanical curve traces, and field transformations all played distinct roles in Catalano's marshaling of geometry in the service of design.[27] For Catalano, these tools remained effectively his private catalog of technique. But his inclinations presaged a much broader and more democratic experimentation with hyperbolic and ruled-surface forms in architecture.

As they migrated from engineering into architecture, warped forms assumed new intellectual and symbolic roles. The engineers who pioneered hyperbolic shapes were industrial entrepreneurs as much as designers, applying their geometric expertise toward iteratively repeatable services and products. Intellectual property protection was a natural market advantage. Unlike the engineers that preceded them, the architects Nowicki, Catalano, and their successors generally did not patent work related to their geometric research.[28] On the contrary, architects were attracted to this new geometry not for its potential as commercial product but rather for its exceptional visual presence. While engineers deployed warped surfaces in factories, warehouses, and markets, architects elevated them to more singular use in churches, assembly halls, and government spaces. At the same time, the mathematical origins of warped surfaces endowed them with a generic flexibility, apparently ahistorical yet enjoying the Platonic qualities of pure geometry. New architectural surfaces became a theater in which the cultural investiture of mathematics played out.

If the shapes of hyperbolic form were geometrically pliable, they were also promiscuously adaptable to the conceptual and ideological proclivities of a wide swath of architects, from iconoclastic experimentalists to the doyens of the International Style. While a few experimental designers, including Candela, Nowicki, and Catalano, plumbed the theoretical, structural, and formal depths of hyperbolic shells in this early experimental phase, eventually the unusual properties of these forms began to command the attention of many of the best-known architects of the 1950s and 1960s. The quasi-nationalistic competition of the earlier engineers was replaced with an unselfconscious borrowing, quotation, and remixing. This proliferation constituted an imperial period of hyperbolic form, during which Marcel Breuer, Hugh Stubbins, Eero Saarinen, and even Le Corbusier adopted the language as their own, placing a canonical imprimatur on the idiom of hyperbolic forms. With that broader attention came a concomitant diffusion of the technical knowledge necessary to achieve new geometries.

Ruled-surface and hyperbolic architecture were felicitously suited to the vaulted halls of public building types. Their lyrical but difficult-to-partition spans were no impediment to the more open volumes of assembly halls, libraries, stadiums, convention centers, churches, and showrooms. The trend toward these new pseudo-gothic forms was notably international, with examples dotting the globe. Japanese architect Kenzo Tange took the hyperbolic projects to the scale of civic structures in Japan, for instance in the Yoyogi Olympic Gymnasium (1964) in Tokyo. Projects like American architect Hugh Stubbins's 1957 Kongresshalle in Berlin (now the Haus der Kulturen der Welt) brought the saddle structure to the midcentury exhibition hall. The four hyperbolic sheets reaching skyward of I. M. Pei's Luce Memorial Chapel (1963), on the campus of Tunghai University, Taiwan, frame a linear skylight that floods the space. Marcel Breuer's Church of Saint Francis de Sales (1967) in Muskegon, Michigan, remixed hypar geometry not for roofs but for mammoth concrete walls. And the unintentionally twinned Cathedral of Saint Mary (1964, by Kenzo Tange) in Tokyo, Japan, and Cathedral of Saint Mary of the Assumption (1971, by the team of Lee, Ryan, McSweeny, Nervi, and Belluschi) in San Francisco, California, create cruciform spaces that interpolate between roof and wall with thin-shell ruled surfaces. Once-extreme warped forms were suddenly a new international language of singular public spaces.

fig. 9.6

Hyperbolic geometries were even embraced by some of the elder statesmen of the moment. While Robin Evans has noted the use of ruled surfaces in at least ten of Le Corbusier's projects, the office undertook a trio of particularly significant designs that drew heavily on that visual language.[29] Though we have already mentioned the Church of Saint-Pierre, in Firminy, France, and the Philips Pavilion in Brussels, Belgium, undoubtedly Le Corbusier's best known ruled-surface project was Notre-Dame du Haut in Ronchamp, France (1954). The elemental but dynamic volumes of this chapel are perched on a hilltop, dramatically overlooking the valley below. Geometrically, the roof is not precisely hyperbolic but instead a conoidal ruled surface generated by lines spanning between a straight edge and an opposing parabolic arch at proportional points. The working drawings of Notre-Dame du Haut make explicit this construction.

fig. 9.7

Not confined to large gestures, ruled geometries were also deployed at more modest architectural scales to dramatic effect. One particularly striking example was Eero Saarinen's staircase for GM's Research Building (1955) in Warren, Michigan.[30] The staircase was suspended by an intricate series of tension and compression rods that delineated a network of virtual ruled surfaces. Diaphanous and transparent, it evokes both Shukhov's intricate towers and the ruled-surface models of Monge and Olivier at full scale. In fully realizing the sculptural quality of this staircase, Saarinen embraced the operative potential of the ruled surface as

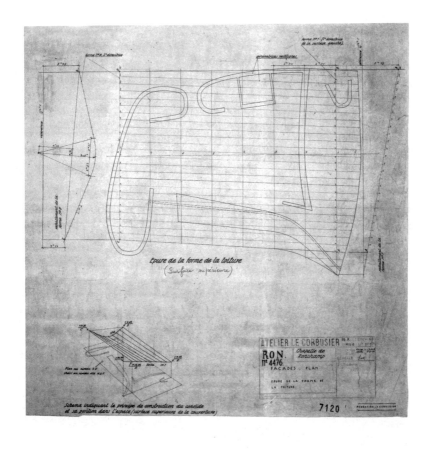

9.6   Drawing of the roof of Notre-Dame du Haut, detailing the ruled-surface conoid structure. Fondation Le Corbusier. © F.L.C. / ADAGP, Paris / Artists Rights Society (ARS), New York 2020.

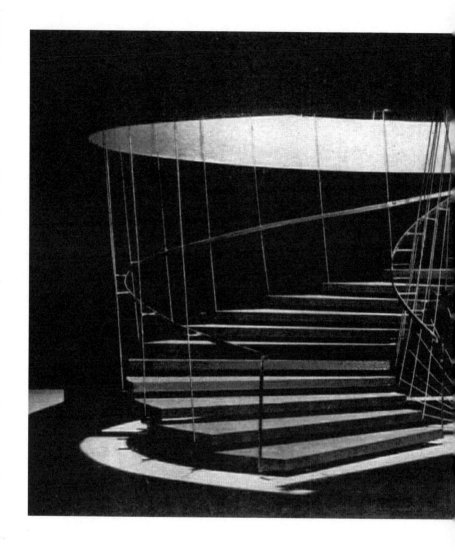

9.7   The main staircase of Eero Saarinen's GM's Research Building.
Source: *Progressive Architecture* 36, no. 2 (February 1955): 102.

a functional and structural element of architecture, not merely a signifier of modernity.

As ruled surfaces inflected the vocabulary of midcentury architecture, they became instantly recognizable emblems of the rapprochement of engineering and design. They were also imminently photogenic, their warped contours washed with light and evoking levitating and weightless forms. The popular press in the United States and Europe took note. Agreeing that "the traditional roof . . . is in some respects an anachronism,"[31] journalists were struck that "the strange shape takes care of the structural stresses."[32] These new forms were "highly contemporary"[33] and "capable of producing a completely new kind of architecture."[34] The press embraced hyperbolic geometry as a symbol of technological advancement and as the shape of things to come.

Particularly in the United States, hyperbolic geometry also inexorably gained currency in the rapidly expanding mainstream architecture press. Circulating in newspapers, popular magazines, and professional journals like *Progressive Architecture* or *Architectural Record*, it became an irresistible signal of an experimental vanguard. The means of communicating geometric knowledge to achieve these forms evolved with the age: broader circulation of images, technical articles, and discussion of hyperbolic structures, particularly in ephemera such as magazines, created an informal knowledge economy that unlocked access to complex geometry for steadily wider audiences. No longer the purview of specialized engineers, cutting-edge geometry became available for experimentation by designers and even hobbyists.

The Marketing Media of Complex Geometry

From the advent of the magazine format in the nineteenth century, architects have recognized the inseparable association between self-promotion and the promotion of industrial products. Every magazine was, in reality, a catalog of advertised goods, and the savviest designers embraced the image as a commercial currency. Beatriz Colomina has noted, for example, "Le Corbusier's use of mass media culture, of the everyday images of the press, industrial publicity, of department store mail order catalogues and advertisements" in promiscuous and inventive ways, either as sourcebooks of ready-made architectural forms or as prototype advertisements to be placed in his own magazine *L'Esprit Nouveau*.[35] In the case of magazines with a specifically architectural focus, the symbiosis between content and product was even more intimate. Unlike the brochures of industrial objects raided for collage elements, in professional magazines architecture itself was content for publicity. The anatomy of architecture was dissected and sold for parts.

The high point for hyperbolic architecture corresponded with a golden age of published trade magazines, and more particularly with the advertising publicity that appeared in those magazines. Product manufacturers seized on the powerful semiotic potential of new geometries as signifiers of a progressive future. A few more innovative companies even directly sponsored the construction of daring prototypes as brand promotions. Sponsored by the West Coast Lumberman's association, John Storr's Forest Products Pavilion built for the Oregon Centennial of 1959 arranged a complex of hypars in a dramatic pavilion roof. An eight-page technical manual, *Simple Hyperbolic Paraboloid Shells of West Coast Lumber*, accompanied the project to "emphasize the ease with which such a structure is designed and built."[36] Paul Hayden Kirk's 1961 Wood Products Research House in Bellevue, Washington, was designed as both a showcase for plywood products and as a promotion piece for *Living for Young Homemakers* magazine.[37] The methods and techniques for its construction were detailed in *Progressive Architecture*—a symbiosis between popular and professional promotion. In 1957 Catalano himself developed an aluminum doppelgänger of his own house at the request of the Reynolds Metal Company.[38] Mediagenic shells endeared themselves to manufacturers attuned to the power of photographic images.

Many product manufacturers also traded on the use of their products in daring buildings to confer a patina of innovation. Playing on ubiquitous fears of leaky hyperbolic shells, an advertisement for Dupont waterproofing products showcased ten striking roof designs and teased, "Who's Afraid of the Big Bold Roof?" Images of telegenic event spaces such as the sinuous Kodak Pavilion by Ely Kahn and Robert Jacobs for the 1964 New York World's Fair became marketing collateral for many products simultaneously. "Exciting design possibilities become realities," as the concrete manufacturer Pozzolith announced of its use on the Kodak Pavilion, while Aerospace and defense contractor Martin Marietta's construction division eagerly touted its own involvement. Pozzolith also trumpeted the "spectacular roof design" of thin shells for St. Louis's modern new 1956 air terminal by Hellmuth, Yamasaki, and Leinweber (a forerunner of HOK).[39] The same terminal was proclaimed by Hillyard Chemical Company as "Where Tomorrow Begins." Without a hint of irony, Hillyard proposed that "The past 50 years has seen unbelievable progress in the development of new forms, new materials, new harmony of design—and the unveiling of new concepts of floor treatment."[40]

The bumper crop of novel concrete forms was celebrated by trade groups such as the Portland Cement Association, or individual contractors such as Lehigh Cement, who promoted their expertise in the construction of hypar roofs. Not to be limited merely to completed projects, some ambitious manufacturers, such as Universal Atlas Cements, brandished dramatically out-of-scale speculative images of unbuilt (and unbuildable) projects, claiming they were the material of "Tomorrow's Arena." Even pedestrian cedar shingles "add strength" to torquing roofs, and glass panels were trumpeted for their ability to complement complex geometries: "a curving concrete shell pierced with light, with glass by ASG." Companies selling elevators, solid-core doors, and even instant cameras, were all eager to align themselves by association with the imagery of remarkable geometries of a new space age. Even smaller projects, like architect Joseph Salerno's 1963 United Church of Rowayton, in Norwalk, Connecticut, were similarly feted. Across an image of the church, the National Timber Manufacturer's association tempts architects to "Choose Wood . . . and your imagination!"

Few buildings provided quite the convergence of professional and popular publicity bonanza of Saarinen's 1962 TWA Terminal at New York City's Idlewild Airport (later John F. Kennedy Airport). Breathlessly covered by the architectural press, it was also an instant icon in the popular media, with its soaring vaults so evocative of flight. *Life* announced that "the terminal stunningly displays the qualities" of "poetic inventiveness, [and] monumental simplicity."[41] Perhaps the most sublime popular homage to the terminal was a cameo in the sixth episode of the popular television series *The Jetsons* in 1962, a cartoon set a hundred years in the future.[42] Saarinen was no stranger to hyperbolic forms, having embraced ruled surfaces in projects such as the 1958 David S. Ingalls Hockey Rink. In fact, engineering historian Tyler Sprague has noted that Saarinen and Nowicki were erstwhile collaborators, meeting at the Cranbrook Academy in Bloomfield Hills, Michigan, when Nowicki was a visiting professor there in 1949.[43] Saarinen acknowledged the significant influence of conversations with Nowicki, suggesting that a kind of geometric folklore may have laid the groundwork for the TWA Terminal.[44]

The advertising ephemera of product manufacturers occupied a zone between public reception of architecture and its professional production. Ads played up the final public experience of the building in broad terms which even nonarchitects could readily embrace. Product marketing amplified a wave of popular and public fascination with these new forms. Commercial exposure complemented a more intense conversation about these projects within architectural circles. As complex geometry was increasingly diffused in wide-reaching professional magazines and journals, the professional media communicated not only what the future could look like, but the geometric training and acculturation necessary to draw it.

Broadcasting Geometry

If these projects were among the most visible public manifestations of a growing interest in the mathematics of warped geometry, their production was underwritten by efforts to disseminate, through the architectural press, practical knowledge on ruled and complex surfaces. Professional magazines became mass-media tutors in the craft of esoteric mathematics. Between 1950 and 1970, serialized articles in mainstream publications demystified the intricacies of hyperbolic, geodesic, and ruled-surface geometries. Sharing the techniques of geometrical complexity became a kind of evangelism, a means of extending the knowledge culture of technical design to any who might be inclined to engage it.

The tutorials provided by *Architectural Record* and *Progressive Architecture* were part of a much broader trend toward accessible education through correspondence courses and mail-order training in the United States during the 1950s and 1960s. Correspondence courses in architectural drafting had been available since at least the 1920s, from both established institutions like Columbia University and for-profit companies.[45] By 1950, the National Home Study Council, a trade group, boasted that such courses taught "an unlimited variety of subjects, from architecture to zoology."[46] Technical and engineering subjects comprised the majority of instruction. The promotional material for these courses noted that "correspondence schools are sure the demand for technical knowledge will grow. They are banking on the premise that the next decade will be one of rapid technological advance."[47] Disciplinary publications like *Architectural Record* used the same basic model of distance learning by guides in fostering more abstract and geometric knowledge to their readership.

Open geometric tutorials replaced elite training with freely available toolkits. The privilege of specialized expertise no longer furnished the sole access to complex architectural geometries. Sophisticated mathematical methods were now encapsulated and circulated in universally accessible guides that distilled geometry to its operative essence.

In 1954, *Architectural Record* began a series of articles on the geometric and technical details of thin shells, asserting "it behooves the American designer to get acquainted with this type of structure in its most efficient forms."[48] Subsequent articles quickly expanded to a broader range of geometric topics: in August 1955, *Record* began a more formal series by Pratt professor Seymour Howard on "Useful Curves and Curved Surfaces," which accessibly introduced many of the touchstone geometric concepts of this time to a wider audience of architects. According to Howard, his motivation was to radically expand access: "Many good designs have never been carried out because information has not been readily available on curve characteristics and methods for laying them out. These and subsequent sheets will provide such information, not only on the familiar curves, but also on curves used for geodesic surfaces and thin shells."[49]

These articles were intended not for a specialist audience but for a broad swath of architects who might adapt altogether pedestrian construction methods to achieve the remarkable geometric results. Mirroring the precision of earlier patent summaries but without the restrictive legal constraints, they were manuals for the self-made geometer. Nevertheless, the topics were introduced rigorously: curves were presented by their parametric equations, standard form, and constructed drawing methods. In all, over thirty specific curves, curved surfaces, or polyhedral geometries were cataloged, including trigonometric curves as well as more advanced cycloids and trochoids. Even more ambitiously, Howard introduced notions from differential geometry, such as Gaussian curvature, minimal surfaces, and catenary curves. He conveyed sophisticated analytic facts, for instance that the sphere was area minimizing and volume maximizing while the tetrahedron was volume minimizing and area maximizing. The articles were mathematical manuals in miniature, crafted for an architectural reader.

Articles like "Useful Curves and Curved Surfaces" broadcast fig. 9.8 the esoteric craft of geometry to the whole discipline of architecture. Many articles responded to particular contemporary projects, decoding their otherwise arcane methods and ranging beyond purely hyperbolic forms to other eccentric geometries. In 1958 Howard provided detailed instructions on the partitioning and construction of icosahedral geodesic domes, in response to the completion of Buckminster Fuller's Kaiser Dome in Honolulu, Hawaii, the year before.[50] In the February 1959 issue Howard particularly treated hyperbolic paraboloid structures, fastidiously describing their characteristic properties by way of classical descriptive geometry.[51] Howard's training extended from the mind to the hand, offering tactile drafting-table methods of drawing these complex curves: "For drawing curves which do not lend themselves to simple mathematical analysis, the best technique is to use wood splines and battens, held in position by lead weights called ducks."[52] This method of approximating mathematical curvature with bent wood and weights was precisely the technique of curve drawing that was used in aircraft, automotive, and boat construction factories, and which shaped the early digital representation of curves.

Howard's work was presented under the rubric of "Time-Saver Standards," and thus the range of geometric technique that Howard introduced in *Architectural Record* was essentially canonized as relevant disciplinary knowledge for architects. Released serially in *Architectural Record* since 1938

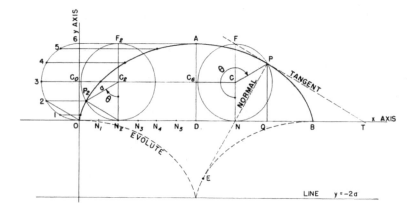

9.8   Seymour Howard's drawing of a cycloid.
Source: Seymour Howard, "Useful Curves and Curved Surfaces: 10—Cycloid,"
Architectural Record 120, no. 3 (September 1956): 287.

and anthologized in book form since at least 1942, Time-Saver Standards functioned as a codex of accepted disciplinary technical knowledge and a common point of reference between architects and building engineers. Like the wider content of Time-Saver Standards, Howard's tone of instruction was matter-of-fact, with little editorializing. In aggregate, the articles were very much in the nature of a manual, laying out the knowledge necessary for a new generation of geometric designers. The pragmatic need for a sourcebook of geometric content is almost taken for granted, as if the complex mathematical subdivisions or intricately curved surfaces were the everyday bread and butter of American architects.

Architectural Record was by no means alone in the effort to popularize complex geometry. Other periodicals, such as *Progressive Architecture*, also acted as organs for the dissemination of technical knowledge related to the frontiers of formal experiment. In March 1956 *Progressive Architecture* devoted an issue to tension roofs, highlighting projects like Nowicki's Dorton Arena and Stubbins's Kongresshalle in detail, with the observation that these new geometries "demand mental training of all of us."[53] These articles invariably introduced simple mathematical rules of thumb to guide the development of new projects, as well as taxonomic catalogs of the modes in which a particular geometric language might be adapted. Articles in *Progressive Architecture* elaborated on the geometry of spheroid grid domes, hypar aggregations, and more. For instance, Edward Tuttle's 1971 piece for *Progressive Architecture*, "Hypar Gambits," instructed designers how to aggregate parabolic modules to enclose specific spatial typologies.[54] Each new formal system became a matter of technique which, with appropriate training, anyone could master.

For more than fifteen years, during the apex of the popularity of hyperbolic form, architectural mass media thus propagated an increasingly egalitarian training for a generation of designers in the pragmatics of complex surfaces and architectural geometry. Series like "Time-Saver Standards: Useful Curves and Curved Surfaces" confirmed the commitment of disciplinary media to rigorous yet consumable mathematics. The audience for once-arcane geometric knowledge was expanding as never before.

Hypar Spaces, Hippie Geometries

While it did not endure as an architectural idiom, the hyperbolic moment provided evidence of the remarkable power of encapsulated mathematical knowledge, and of the attraction of using democratic means to distribute complex technical procedures. Though there were inevitably avatars of the formal paradigm—the Candelas or the Saarinens—there were far more independent practitioners who entrepreneurially and anonymously developed hypar and thin-shell projects which dotted the byways of mid-twentieth-century America and Europe. Paradoxically, the mass media of popular and trade publications shaped the cultural image of advanced geometry as both avant-garde and democratic. This duality of advanced futurism and grassroots, even hippie, ideals would prove to be an unexpected but enduring dimension of the culture of architectural geometry.

fig. 9.9

Remarkable as this explosion of warped shells proved to be, the form also had stubborn architectural limits. Even evangelists observed that the formal integrity of these shells—at first an elegant attraction—proved so relentless as to subvert attempts to partition, subdivide, or otherwise intervene in its strict logic. Labor, also, was a limiting factor: while thin shells minimized material, extraordinary work was required to properly craft all of their customized pieces. The most prolific shell builders were often practitioners in geographies with low labor costs relative to material costs, such as Central and South America. Whether because of these formal limitations or a cultural exhaustion, 1970 proved to be a turning point, and the large-scale experimentation with hyperbolic forms seemed to dissipate as abruptly as it had emerged. Mentions of hypar geometry in print media declined precipitously, and architectural interest virtually evaporated. The hypar was thus banished to the purgatory of the perpetual future. Perhaps because it did not quite succeed in becoming ubiquitous to the point of vernacular, it persisted as a sign of unrealized architectural potential, a future tantalizingly out of reach. Thus in 1980, well after the hypar heyday, *Popular Science* could still claim that the hypar was a "unique and revolutionary building system, a system that might just dramatically change the shape of our

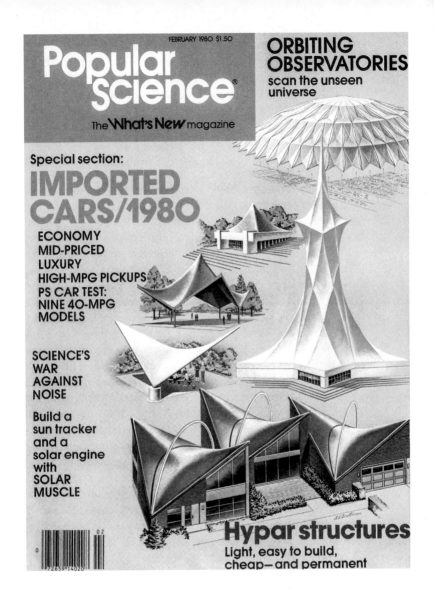

FEBRUARY 1980 $1.50

Popular Science®

The **What's New** magazine

ORBITING OBSERVATORIES scan the unseen universe

Special section:

IMPORTED CARS/1980

**ECONOMY
MID-PRICED
LUXURY
HIGH-MPG PICKUPS
PS CAR TEST:
NINE 40-MPG
MODELS**

**SCIENCE'S
WAR
AGAINST
NOISE**

**Build a
sun tracker
and a
solar engine
with
SOLAR
MUSCLE**

Hypar structures

Light, easy to build, cheap—and permanent

9.9 The cover of the February 1980 issue of *Popular Science* announces the sundry advantages of hypar architecture. Source: *Popular Science* 216, no. 2 (February 1980). Used with permission of Popular Science, © 2020. All rights reserved.

homes, churches, schools, and offices within a few decades."[55] Warped geometry was a perpetual mirage on tomorrow's horizon.

The proliferation of hyperbolic geometry in architecture cuts an identifiable path of mathematical techniques through the broadest reaches of the profession. In the arc of the hyperbolic moment, there was a specific social progression: from a technically esoteric cadre of engineers, to a looser network of architectural collaborators, to an open, public, and prolific promotion through the new professional mass media and cross-sponsorships, and finally to a point of evident oversaturation and decline. This arc had all the hallmarks of the lifecycle of a distinct and coherent architectural subculture.

Despite their near-disappearance from architecture, hyperbolic forms whetted the architectural appetite for ever more daring geometries and experimental techniques. Thus, while development of hyperbolic forms per se seems to abate after 1970, fascination with complex geometry more generally did not subside. On the contrary, if anything, 1970 brought a still broader expansion of geometric vocabularies for the architect and builder. The open, democratic, do-it-yourself ethos accelerated with the advent of even cheaper and more radical structures such as inflatables, geodesic domes, and zonohedral structures. We have already seen the remarkable *Cubic Constructions Compendium* released in 1970, which aspired to be a free and open sourcebook of voxel architecture. Dome fever was also in full swing then, and as the geometric varieties of domes multiplied, so did the media for consuming knowledge of their intricate geometry.

Often ideologically at odds with the commercial motives of larger magazines, the proponents of these new architectures nevertheless saw value in the development of manuals of form to train willing designers in the newly accessible arts of complex geometry. From this moment emerged a new type of media: the DIY geometry zine, a tacit rebuttal of the commodification of geometry in the professional press. The prototype was Stewart Brand's *Whole Earth Catalog*, first released in 1968. Self-published as a clearinghouse of resources for independent thinking and making, its subtitle, "Access to Tools," neatly summarized the anti-elitist ambition toward encapsulated knowledge.[56] Self-produced and entrepreneurially distributed, often by word of mouth, the DIY zine fused technical geometry with a playful and open approach to knowledge culture. Several notable self-published manuals on idiosyncratic but geometrically provocative construction methods were released around 1970 that took a more egalitarian approach to design technique to its logical extreme.

Geodesic polyhedral structures were indelibly registered in the popular media not only by the structures of Buckminster Fuller but also by the Fulleresque colonies of domes built from recycled car bodies that came to be known as Drop City. Drop City was an artist collective founded in southern Colorado in 1965 with the aim of providing a free and nonhierarchical community for the practice of art. A beacon of hippie culture, Drop City also created an image of what a progressive intentional community should look like. Architectural experimentalist Steve Baer's *Dome Cookbook* became the indispensable manual for do-it-yourself construction of domes like those of Drop City builders.[57] The *Dome Cookbook* was a curious hybrid: a condensed manual of polytopic geometry combined with pragmatic guidance in the building of nonstandard wood frames. Using the torn-off tops of wrecked cars, Drop Citizens built sophisticated polyhedral structures from Baer's instructions. Baer, who had briefly ventured to Zurich to study mathematics at Eidgenössische Technische Hochschule Zürich (ETH), conveyed a method that was at once rigorous, idiosyncratic, and irreverently ad hoc. Covering everything from arcane aspects of crystallographic geometry to assembly details using common hardware, Baer conveyed an approachable but exact intensity that made relatively advanced geometric ideas accessible for more quotidian construction. Naturally, the *Dome Cookbook* was itself featured in the *Whole Earth Catalog*.[58]

The *Inflatocookbook* did for inflatable structures what the *Dome Cookbook* did for zonohedral domes. Born of San Francisco's countercultural ethos and first self-published in 1971 by the architectural collective Ant Farm, this comic book crossed with a manual laid out in accessible form the engineering content necessary to develop daring inflatable structures. Explaining assembly and geometric parameters in an avuncular manner straight out of *Popular Mechanics*, with crude hand-drawn cartoons and humorous asides interspersed with assembly diagrams and data sheets, it achieved cult status among radical designers of the period. At a certain level, the *Inflatocookbook* provided all of the ready-made ingredients necessary for a new architecture: "With these new Ant Farm components you can now realize your fantasies with most of the dirty work done already."[59]

Manuals like the *Inflatocookbook* and especially the *Dome Cookbook* were a radical endgame of decades of progressively more democratic attempts to bring geometric, mathematical, and technical knowledge to the people. The impulse toward ever broader and more consumable formats of geometric technique that began with the capsule tutorials in *Architectural Record* on hyperbolic geometry ultimately drove an entire literature of democratized geometric design. Channeling a sensibility in the broader popular culture which embraced decentralized modes of education and action, architectural zines and the mathematical and geometric knowledge they contained became vehicles to short-circuit the received arbiters of architecture. Zines reinforced emerging subcultures of geometric production. The sociologist Michelle Kempson notes that zines participate in a "networked practice of exchanging self-produced cultural artefacts, but also … an emerging mode of collective identification for many people invested in the idea and practice of autonomous cultural production."[60] The viral zine was a means not only of education but also identification.

While geometric architectural zines aimed to circumvent institutions, they also empowered their producers as architects: they created channels for the amplification of hitherto-idiosyncratic design experiments. Zines communicated knowledge and formed a culture around that knowledge. The mediatization of design geometry that began with hypar architecture ultimately triggered a revolutionary fervor and countercultural drive that inflected the conversation about mathematics in design. As designers created their own viral and democratic mass media, a feeling gained traction that geometric systems would change not only what we built but how we lived. The lines between the esoteric mathematical culture and mediatized publicity became irrevocably blurred as the new geometries of tomorrow came into focus.

10   Crystal Collectives: Architecture's Chemical Subcultures

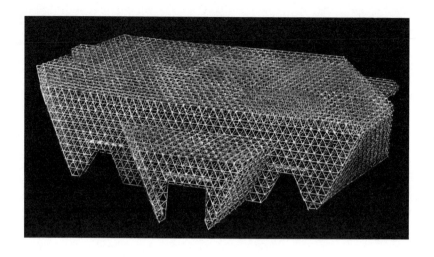

10.1   Synergetics, the firm of Buckminster Fuller, airplane hangar developed with a crystal
lattice-based tetrahedral structure, 1955.   Source: Collection of the Museum of Modern Art,
Object Number MC 28. Digital Image © The Museum of Modern Art/Licensed by SCALA /
Art Resource, NY. Courtesy The Estate of R. Buckminster Fuller.

fig. 10.1

"Today it would be more profitable for a young architect to spend six months in a crystallographer's laboratory than to go to Rome to make some beautiful renderings of ancient monuments."[1] With this bold pronouncement, French architect-engineer Robert Le Ricolais (1894–1977) offered crystallography as a new paradigm of mathematized design. This strange synthesis of the technological, mineral, and biological was cast as a new universal language of transscalar design, a vision of form that was both antinatural and hypernatural. An eclectic cohort of visual experimentalists and crystallographically inclined architects sensed the prodigious implications of mathematical chemistry, from the galactic to the atomic. At its height, crystallography was an entire design cosmology, and a peculiar new cellular architecture was its singular manifestation.

Propelled by Cold War educational reforms, scientific popularizations, mathematized drafting techniques, and new algebraic methods, a group of mid-twentieth-century designers seized on crystal structure as a means to make cybernetic, environmental, and material behaviors interoperable and visible. While analogical associations of crystal form beguiled modern architects since at least the 1920s, what distinguished this new iteration of crystal mania was an increasingly rigorous methodology that drew on mathematical notions of algebraic group theory, homomorphism, and lattice structuralism. Floor plans adopted the qualities of molecular diagrams, and by the 1970s, crystallographic ideas reformulated how architecture could be generated along quasi-algorithmic lines. In the vertiginous prism of the crystal lattice, architects glimpsed a technique for the exhaustive partitioning of modular space and a total science of shape.

The Social Matrix of Crystal Form

Enterprising architects in the United States and Europe espoused a menagerie of alternative and countervailing adaptations of crystallographic architecture. A rogues' gallery of designers including Anne Tyng, Steve Baer, and Peter Pearce ranged intellectually from arcane pseudo-mystic conflations to countercultural subversions of canonical mathematical methods. Clustered at first into small networks of interlocutors, designers increasingly developed conduits for researching and diffusing crystallographic processes that drew them into more organized conversation. Designers themselves were not the only catalysts: individual initiative was often serendipitously amplified by design schools that promoted hard science models of research. At its fever pitch around 1970, the crystallographic tendency coalesced into an insular subculture, with specific diction, visual conventions, institutionalized organizations, and characteristic social norms. Research groups and pedagogical programs developed around this international and multifaceted epistemic tendency, guided by texts and catalogs of crystal form that encapsulated conventions of representation and drawing. Throughout the 1960s and 1970s, an identifiable strain of building project rendered these speculative ideas in tangible form. But the architecture of chemical crystallography also reached beyond buildings to fuse biology, mineralogy, physics, and mathematics in an eerie threshold between animate and inanimate form. In the crystal matrix, static turned fluid, geology intersected psychology, and cold scientific methods found new resonance with living architecture. Not content with architecture alone, designers used crystallography to design a whole philosophy of reality.

The sociology of crystal architecture is limned by a constellation of interconnected sources ranging from the authoritative, like the citation networks of core texts and conferences, to the more ephemeral, like oral histories, unpublished manuscripts, and folklore of the living protagonists. Coded signs of cultural assimilation, such as specific mathematical vocabulary or theorems in these texts, also betray the very particular affinities of these architects. These clues expose not only personal sympathies within an architectural clique, but also orientations toward external intellectual disciplines and interlocutors. Specific projects, drawings, and models unveil distinct techniques and characteristic formal conventions. In concert, these sources form the lattice of crystal sociologies.

The peculiar potency of the crystal clique raises questions related to the acculturation and diffusion of technical knowledge not only within architecture but across creative and exact disciplines. Social dynamics played an essential role. Of particular relevance is the Polish biologist and philosopher Ludwik Fleck's notion of the "thought collective" or "thought group":

> people exist who can communicate with each other, i.e. who somehow think similarly, belong, so to say, to the same thought-group. . . . Scientists, philologists, theologians, or cabbalists can perfectly communicate with each other within the limits of their collectives, but the communication between a physicist and a philologist is difficult, between a physicist and a theologian very difficult, and between a physicist and a cabbalist or mystic impossible.[2]

Fleck's observation is doubly relevant: first, it identifies and anticipates the fractured communication between thought collectives of designers and thought collectives of scientists in our study, and second, it offers a structural term for the technically oriented subgroups within design that emerge around specific scientific modes of thought. In the case of architecture and crystallography—or architecture and any other variety of mathematics—there are actually two thought collectives in mutual interaction. By qualifying the gradient of inclusion in a knowledge culture, Fleck's notion of a collective also begins to suggest a topology of relationships among cultures that qualifies their permeability and mutual affinity. Today, as design promiscuously intersects with other fields, the category of the thought collective—and its appearance in the genesis of crystallographic architecture—offers a clue to the lives and deaths of design subcultures.

Of Minerals and Animals

As a conceptual category, the crystal is uncannily alive, a weird body with mineral and biological qualities. Properly initiated architects of the mid-twentieth century saw crystals as abstract relational configurations of matter by which the complexities of both genetic and static material were distilled to interatomic position and angular measures. British sociologist and futurist John McHale (1922–1978), a frequent collaborator with Buckminster Fuller, wrote in 1968 that crystal form was a watershed with implications "from unravelling of the micro lifecode at the molecular

fig. 10.2

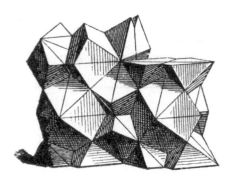

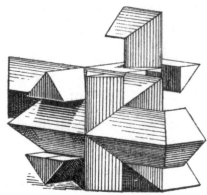

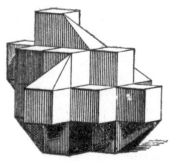

10.2   Three examples of cellular lattice models.
Source: Walther Dyck, *Katalog mathematischer und mathematisch-physikalischer Modelle,
Apparate und Instrumente* (Munich: C. Wolf & Sohn, 1892), 251.

level, to the successful maintenance of men beyond the earth's atmosphere and under its oceans and the outward monitoring of other worlds and galaxies."[3] Crystal form was nothing less than the code of the cosmos itself.

Crystal lattices are the geometric arrangements that organize molecules into regular and repeating spatial networks. These structures originate with a polyhedral unit cell—a simple form like a cube, tetrahedron, or octahedron—infinitely repeated and translated in space along specific linear axes. Replicated unit cells induce a limitless, symmetric, modular lattice that propagates itself universally across the expanse of space.

fig. 10.3

The shape of crystal form first lodged itself in scientific and public consciousness through mineralogical models that jostled alongside wire and plaster maquettes in the shelves of fin-de-siècle mathematical *Wunderkammern*. Extractions of infinite cellular aggregates, they rendered legible the rules of kaleidoscopic prisms and lattices which captivated mineralogists. The tectonic maquettes followed from new understandings of the geometry of crystal aggregations developed by French mineralogist René-Just Haüy (1743–1822). Haüy's work on mineral geometry, particularly *Traité de minéralogie* (1801) and *Traité de cristallographie* (1822), articulated the periodic structures of crystal shapes. Mathematically, Haüy formulated the shear plane properties of crystal aggregations as well as lattice-based laws of crystal growth.[4] His work was also emblematic of the representational turn taken by the sciences after 1800. Beguiling wireframe drawings, translated into a bevy of beautiful physical models, marked a new era of quantized scientific visualization. A generation of crystallographers followed Haüy in ever more intricate and nuanced methods of drawing, modeling, and combinatorially classifying structural form.

Even at this early moment there was a strong thematic rapport between this new science of crystal modularity and the cellular work of certain nineteenth-century architects, such as the French architect Jean-Nicolas-Louis Durand (1760–1834). Durand's ambition was to develop a framework not only to design specific buildings but to unveil the logic of the "composition of all buildings," facilitated by the recombination of architectural elements such as walls, doors, or windows in regular cellular armatures.[5] As the historian Jacques Guillerme has observed, "A close

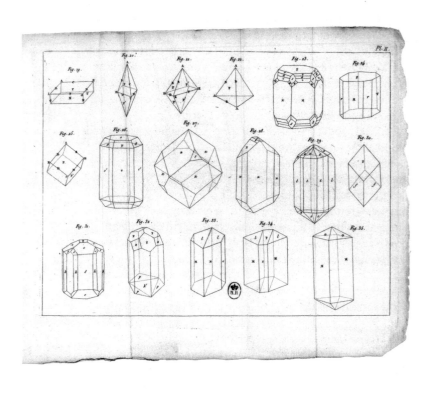

10.3   Diagrams of discretized mineral structures from Haüy's *Traité de minéralogie*.
Source: René-Just Haüy, *Tableau comparatif des résultats de la cristallographie et de l'analyse chimique: relativement à la classification des minéraux* (Paris, 1809), pl. II.
Bibliothèque nationale de France.

reading of crystallographic texts reveals all sorts of indications of the isomorphism of structures which define the 'mechanism of the formation of crystals' and the cells which are at work in Durand's architectural 'mechanism of composition.'"[6] Indeed, isomorphism, isometry, and homomorphism become key variants of the symmetries that underlay crystal architecture. Guillerme even draws parallels to the incipient use of graph paper as a modular composition tool par excellence, later adopted by crystallographically inclined designers to enforce isometry.[7] The geometric regularity and exactness of lattice forms proved irresistible to the rationalist sensibilities of the late Enlightenment.

The discovery of pulsating and evanescent liquid crystals upended intuitions of static crystalline structure embodied in Haüy's drawings. German physicist Otto Lehmann revealed the strangely effervescent and visually arresting patterns of crystal flow in his 1904 work on liquid crystals, *Flüssige Krystalle*.[8] Neither object nor fluid, liquid crystals constituted a matter of perpetual phase change. His 1921 sequel, *Flüssige Kristalle und ihr scheinbares Leben: Forschungsergebnisse dargestellt in einem Kinofilm* (Liquid crystals and their apparent life: research results represented in a cinema film), made biological associations even more explicit. Perhaps most striking was that these crystals exhibited none of the clear reticulation of their more obviously mineralogical counterparts. The historian Spyros Papapetros observes,

> Their structure had nothing to do with the hexagonal snowflakes and other symmetrical polyhedra of the nineteenth century, such as the regular architectonic formations we are used to seeing in the books of Semper, Ruskin, and Viollet-le-Duc. In Lehmann's microphotographs, the new crystals appeared flowing and circular, producing ambient light effects. They formed complex spider webs, or were filled with oil-like patches expanding in a mucus substance.[9]

Through unique images of microscopy developed through a custom camera, crystal form felt mutable, boundlessly elastic, and alive. Suddenly crystal and animal did not feel so distant, but suffused with the same vital energy.

The intricate and turbulent geometry of liquid crystals resonated with the dynamic and modern mathematics of complex surfaces. As the literary critic Esther Leslie has shown, Lehmann cited the Belgian physicist Joseph Plateau's suspended oil film experiments and drew inspiration from the emergent geometry of minimal surfaces.[10] The surfaces of oil films and liquid crystals shared a shimmering iridescent quality, but also a logic of bounded frames. Liquid crystals were not entirely amorphous, but often locally displayed a field structure akin to these new non-Euclidean geometries. They tended to organize themselves in fractured semipatterns of singularities that evoked magnetic fields. Ernst Haeckel, best known for his lavishly illustrated biological works, was profoundly moved by Lehmann's research into the lively animism of liquid crystals. In his 1917 *Kristallseelen: Studien über das anorganische Leben* (Crystal souls: studies of inorganic life), Haeckel argued for the study of crystal form as a metadiscipline, binding together "crystallography and mineralogy; physics and chemistry; morphology and physiology; zoology and botany; psychology and mathematics."[11] For Haeckel, Lehmann's research cracked the door open to a bizarre new synthesis of dead matter and living souls, the possibility of unifying the "crystal and psyche" in a new monistic superscience for the twentieth century.

Crystal Intuition and the Modern Turn

Early modernists were seduced by the promise of chemistry, and its structured corollary crystallography, to lift the curtain from the hidden logic of the material world and endow modern life with a new dynamism.[12] Their arguments overflow with rhetorical or analogical references to crystal structure. For instance, Amédée Ozenfant and Le Corbusier's 1924 essay "Vers le cristal" argued that in the crystal, "nature shows us the way that it constructs its forms, by the reciprocal interplay of internal and external forces."[13] László Moholy-Nagy espoused a more atomic view, proclaiming the crystal as one of seven quasi-elements of design. He enthused: "The biologist Raoul Francé has distinguished seven biotechnical constructional elements: crystal, sphere, cone, plate, strip, rod, and spiral, and says that these are the basic technical elements of the whole world."[14] Yet most significant for our purposes was a tentative but emerging intuition of the operative use of crystal structures in the arguments of Theo van Doesburg in the 1920s.

Van Doesburg saw crystal geometry as a tool for the reconstitution of culture and the establishment of a common format of design communication. As the historian Linda Dalrymple Henderson has shown, he formalized the avant-garde fascination with non-Euclidean geometry and hyperdimensional crystallography by insisting that only these new geometric systems could transcend the spatial conceptions of the past. Through a series of essays in his journal *De Stijl*, van Doesburg offered variations of the hypercube and its derivatives as archetypes of a new kind of crystalline plastic space—folded, projected, or unrolled—and proposed rules for transcribing crystal effects into architecture. Inscribed next to one 1927 drawing of the four-dimensional hypercube crystal, he declared: "a new dimension penetrates our scientific and plastic consciousness."[15] Ironically, the jeweled and glassy bodies of crystals almost seemed to transcend the material realm entirely. Van Doesburg reveled in these contradictory associations, making the crystal a vehicle of both the perfection and transcendence of architecture.

In applying the universal geometry of crystal form to architecture, designers could overcome the vagaries of personal authorship and align themselves with natural laws. For van Doesburg, crystalline geometry was the Trojan horse for a dynamic, universal, objective approach to creation.[16] To the extent that art represented a willful and idiosyncratic vision unresponsive to universal conditions, art itself must be overcome. Or, as van Doesburg stated even more bluntly, "For the sake of progress we must destroy art," and, "The word 'art' no longer means anything. Instead thereof we demand the construction of our surroundings according to creative laws."[17] Ultimately, crystallography went beyond even van Doesburg's objective creative laws to the connection of architecture with nature itself. Through the mathematics of lattice forms, architecture gained access to operable rules of material and biological organization, but also a resonance with some ground truth of the similarly vital experience of modern life. The deep structures of both nature and artifice were coded in crystallographic operations.

353

The aspirational allusions of van Doesburg and his peers gave way, by the 1950s, to more rigorous design methods rooted in crystallographic mathematics. These methods multiplied crystal logic in a hypersymmetry of faceted forms that overcame the limits of more Cartesian conventions of design. Unquestionably, well-known architects like Buckminster Fuller, Louis Kahn, and Robert Le Ricolais promoted crystal structuralism within architecture. But broader cultural currents were also at play. The appearance of influential popularizations of crystallography, which had undergone rapid theoretical and practical development, introduced this arcane science to architects. In 1952 the renowned mathematician Hermann Weyl published *Symmetry*, a widely read anthology of lectures that explained crystallographic notions of rotation, reflection, and translation algebras to a broad audience. Weyl's generously illustrated text was a project in epistemic synthesis that aimed both to "display the great variety of applications of the principle of symmetry in the arts" and to "clarify step by step the philosophico-mathematical significance of the idea of symmetry."[18] Like David Hilbert's *Geometry and the Imagination*, *Symmetry* had an enormous impact in shaping popular intuitions of visual mathematics, as well as opening up the interpretive range of the whole notion of symmetry. Weyl's *Symmetry* became a cultural phenomenon, a "publishing sensation" translated into over fifty languages.[19] Architects from Lionel March to Robin Evans were among its admirers, and Weyl was touted as "essential reading" for geometrically inclined designers.[20]

A further proof of the popular fascination with the crystal was the 1951 Festival of Britain, a manifestation not only of all things British but also of all things crystalline. The festival was a huge national exhibition encompassing over a dozen venues and celebrations, and it became the engine for redevelopment of London's South Bank. The Festival Pattern Group, a committee that coordinated the celebration's official design efforts, trumpeted the intricate patterns of X-ray crystallography as a decorative art for the chemical age.[21] These were symmetries and semisymmetries at the scale of the molecule, transcribed in intricate but consumable graphic form. Almost thirty manufacturers of home goods, furnishings, and ephemera developed graphic designs heralding molecular structures as patterns attuned to modern life. For European designers, it was a carte blanche endorsement of chemical form as a contemporary visual idiom.

Amid this broader moment, a cadre of liminal young fig. 10.4 American and European designers devotedly embraced the bolder and more mathematical possibilities of crystallography, abandoning the symbolism that characterized van Doesburg's use of crystals. For designers like William Huff, Peter Pearce, Arthur Loeb, and Steve Baer, crystal lattices exhibited a powerful set of methods to make spaces discrete and therefore calculable and recombinant. This more thorough embrace of a crystal method came at a point of cultural inflection. Universities and schools suffused with technocratic Cold War tendencies were only too willing to underwrite research into a mathematically rational design. Both designers and institutions entered the glittering world of crystal architecture.

This generation of designers jettisoned facile analogical associations between building and crystal to champion the mathematics of abstract lattice algebras that could more readily compute form. Their resolute commitment to rigor is legible not only in representational techniques but also in their use of specific diction and classification theorems accessible only to those with exposure to discrete mathematics. Indeed, knowledge of each domain of discrete algebra became a distinct level of esoteric initiation, a further register of cultural fluency with specific methods of combinatorial creation. To wit: the recitation of the five Platonic solids, thirteen Archimedean solids, seven frieze groups, seventeen wallpaper groups, and 230 space-filling groups became almost a catechism for architectural texts inspired by crystal form, a necessary part of almost any introduction to architectural geometry. Gratuitous displays of mathematical erudition were tokens of subcultural distinction.

Superscalar Lattices

In 1969, at the zenith of architectural interest in crystal form, a special issue of the Italian periodical *Zodiac* took stock of the state of the art in architectural geometry. It collected essays on crystal

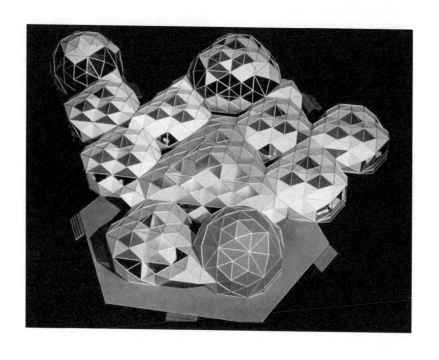

10.4 A proposal indicative of the potential of Peter Pearce's crystalline method of spatial
subdivision and construction.   Source: Peter Pearce, *Structure in Nature Is a Strategy for Design*
(Cambridge, MA: MIT Press, 1978).

form by John McHale, Moshe Safdie, and Anne Tyng, among others. It also juxtaposed Buckminster Fuller's mathematics of rigid polyhedral motions with Austrian architect Frederick Kiesler's research into endless and continuous smoothly topological forms. In her introduction, the editor Maria Bottero assessed the dynamic qualities of Kiesler's work against the more regular forms of Fuller:

> Nothing could, apparently, be more dissimilar than the "endless" fluidity of Kiesler and the severe determinism or geometric structuralism of Fuller. . . . It is, however, just for this reason that a confrontation seems eloquent. In each case, in fact, the starting point is the breaking up of traditional cubic space. . . . Even though along different lines, none the less, matter reigns supreme, both in Kiesler and in Fuller, as an energetic and dynamic system.[22]

Rather than focus on the contrast between Kiesler and Fuller, Bottero stresses their shared theoretical commitment to an underlying logic of physics. Crystal geometries could act as a common multiscalar language of the material environment, microscopic and macroscopic, animate and inanimate, natural and artificial.

Featured prominently in the same publication, the American architect Anne Tyng's (1920–2011) crystallography elided the natural and the artificial in dramatic new transscalar structures in which symmetry took on intensely human, almost psychological associations. An early proponent of crystal approaches in architecture, Tyng was a designer and theorist of geometry, and partner in Louis Kahn's practice for over twenty years. In her essay for *Zodiac*, she extrapolated from the Platonic molecular structure of graphite, salt, or diamond to what she termed "metamorphology," a Felix Klein–like classification of crystal transformations.[23] Tyng illustrated the transscalar expanse of crystal form through a striking comparison: her diptych presentation of the tetrahedral structure of Tyng and Kahn's City Tower in Philadelphia beside the crystalline maquette of Watson and Crick's DNA double helix. Assembled in precisely the same node-and-bar structure of scientific molecular models, and embracing the same crystalline geometries which define molecular form, the City Tower project emanated from an impulse resonant

fig. 10.5

10.5   Louis Kahn and Anne Tyng's City Tower project, which Tyng compared to Watson and Crick's DNA model.   Credit: Louis I. Kahn Collection, University of Pennsylvania and Pennsylvania Historical and Museum Commission.

with the double helix, Tyng argued: "It seems likely that the archetypal images of the unconscious mind were indeed pre-formed by helical energy-form tensions in the creative process of this discovery."[24] The ultimate code of natural and artificial structures was homological with crystal lattices, and through such geometry the inanimate was thereby animated. The architect Moshe Safdie (b. 1938), who befriended Tyng in 1962 while they were both working with Kahn, observed that "she approaches the environment from the atoms, molecules, and crystals that make it up."[25] Through crystal models, Tyng searched for the "architectural DNA molecule"—that which brings life to design.[26] She believed crystal form could collapse geometric and biological structure into a new organic unity.

fig. 10.6

Crystallography was a marvelous tool for the production of spatial difference. Intricate crystal arrangements produced mazes of endless oblique niches, nooks, and articulated rooms, transgressing every canon of the conventional architectural plan. Anne Tyng's proposed dormitory project of 1963 is an instructive example of the genre "conceived," she reported, "as a 'molecular' plan of squares and octagons."[27] The checkerboard of these figures provided "an ordained location for closets, utilities, and, in some cases, are used as sleeping-study alcoves," as well as providing a natural partition of server and served spaces.[28] It was an architecture of rigorous interiority, partitioning space into a filigree of rhyming but differentiated cells. Tyng's dormitory was a mineral for living. Geometric operations like truncation of polyhedral corners become architectural corbelling, creating "Alhambra-like stalactites."[29] Tyng was not alone in her interests, and architecture as crystal aggregation proliferated in academic circles as well. For example, the British architect John Frazer, who would later become a pioneer of computational design, proposed a remarkable series of interlocking staircases in his studio project at the Architectural Association which were, at their core, architecturalized crystals akin to Tyng's molecular spaces.

fig. 10.7

The "New Math" of Design

The virtuous cycle of cultural momentum and exploding design interest provoked young architects to push crystal form in startling new directions. They were encouraged by educational

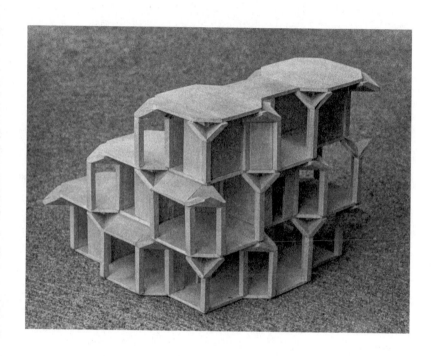

10.6 Anne Tyng's project for a Proposed College Dormitory (1963), a room configuration she described as "molecular." Credit: Anne Griswold Tyng Collection, The Architectural Archives, University of Pennsylvania.

SECTIONAL ELEVATION OF CORE ASSEMBLY

10.7   John Frazer, third-year AA studio project, demonstrating an early interest in crystal packing, 1965.   Source: Architectural Association Archives, London.

institutions and, indirectly, by Cold War governments, which saw scientific rationality as necessary and appropriate to apply in domains as far afield as architecture.[30]

Recasting design around advanced algebraic and geometric topics paralleled the broader "New Math" movement of the late 1950s and 1960s. New Math was an attempt to intensify scientific instruction by introducing set theory, abstract algebra, matrix operations, and other advanced topics to high school and even middle or elementary school students. Mathematics was lauded as a metadiscipline of the sciences that endowed students with a specific kind of mental training. Citizens with disciplined mathematical reasoning were essential to the demands of the Cold War arms buildup and the space race. Originating in the United States in the work of the National Science Foundation–sponsored School Mathematics Study Group (SMSG), New Math reshaped the way a generation of Americans would encounter mathematics. The SMSG proposed several model curricula between 1958 and 1972 that proved profoundly influential not only in America but internationally as well.[31] They inspired European programs such as the UK's School Mathematics Project, which in turn had a parallel impact on British teaching.[32] As the historian Christopher Phillips has observed, New Math's goals were heuristic: "Math class was said to provide epistemological training—teaching students about what counts as valid knowledge and the grounds for its validity."[33]

New Math prioritized an operative perspective of mathematics that deeply resonated with geometrically inclined designers. Architectural texts with a mathematical orientation aimed to rethink design with the New Math: "Our aim is twofold: one, to help bridge the gap between the new mathematics and the older generation; and two, to suggest to the youthful reader, perhaps with a science and mathematics background, that architecture is an exciting subject."[34] The young American designer William Huff (b. 1927) particularly saw the enchanting promise of New Math's crystal algebras for architecture. He was primed from his systematic study of spaceframe and dome geometries during his time at Yale's architecture school, from 1950 to 1952. The rising architect Louis Kahn was Huff's thesis critic, and when Kahn began his own study of crystallography that ultimately shaped the City Tower project, Huff was one of his closest students.[35] After graduation, Huff worked in Kahn's office and no doubt interacted with Anne Tyng during this period as well.[36] Huff ultimately decamped to Ulm's Hochschule für Gestaltung (HfG) in 1956 on a Fulbright scholarship, and in 1963 he began teaching Ulm HfG's Basic Course, giving the exercises a structural and quasi-algorithmic focus.

At that moment, Ulm HfG was an enviable match for Huff's research, acting as an incubator for fertile new methods of exact design like topology and crystallography. The Basic Course in particular was a crucible of the new mathematized design. Earlier instructors of the course included Josef Albers and Hermann von Baravalle (1898–1973), a mathematician who had previously developed the highly visual methods of teaching projective geometry for Rudolf Steiner's Waldorf Schools. Von Baravalle, who began to teach the course in 1955, introduced exotic topics in geometry, including differential subjects such as the kinetic traces of points and lines similar to the light motion studies of scientific management.[37] Tomás Maldonado immediately preceded Huff in the Basic Course, and his version had first introduced topological and symmetry operators.[38] Constant experimentation with the logic of geometric pattern animated the pedagogy of the Basic Course.

fig. 10.8

When Huff assumed leadership of the Basic Course, his teaching studio became a stage for meticulous experimentation with planar and volumetric symmetries as well as mesmerizing spatial patterns. Drawing on a strong knowledge of the history of mathematical crystallography, between 1963 and 1968 Huff produced two species of exercise that were characteristic of this experimental impulse: solid dissections and parquet deformations. Huff formulated a series of design exercises on the dissection and reassembly of cubes and other solids into twofold symmetries, creating odd volumes that seemed to turn these solids inside out. The exercise echoed the stereoisomer work of Pasteur (and were in turn echoed by the later work of Arthur Loeb), and indeed Huff's familiarity with Pasteur's crystal models is evident from his writings. These dissections proposed strange proto-architectures, twinned chambers in a frozen dance around a common axis.[39] If these dissections toyed with isometric symmetries, parquet deformations, Huff's second pedagogical innovation, plumbed

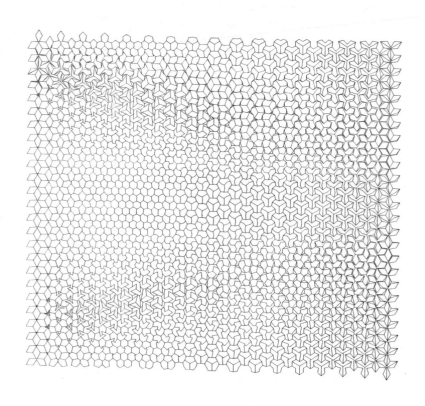

10.8 Example of a hand-drawn crystallographic tiling transformation from William S. Huff's Ulm HfG course, ca. 1965. Source: William S. Huff, "The Parquet Deformations from the Basic Design Studio of William S. Huff at Carnegie-Mellon University, Hochschule für Gestaltung, and State University of New York at Buffalo from 1960 to 1980," unpublished manuscript. Courtesy of William S. Huff.

more differential symmetries through subtle incremental lattice transformations that continuously morphed unit cells into fluid kaleidoscopic effects. They implemented graphically homeomorphic transformations of tile patterns, operations that were at once topological and group-theoretic. Though dizzyingly ornamental, parquet deformations most appealed to Huff due to their visual structure and the suggestion of complex spatial partitions. They were a designed parallel of liquid crystals.

figs. 10.9, 10

The collective of designers coalescing around crystal mathematics in the 1960s was eclectic and ecumenical. While some experimentalists like Huff were self-taught, others, like the physical chemist and design theorist Arthur Loeb (1923–2002), tapped rigorous scientific training to reimagine design methods at their foundations. Loeb was educated as a physical chemist at the University of Pennsylvania and later at Harvard, where he completed his PhD in 1949. For Loeb, crystallographic lattices unlocked invariant principles of spatial structure, visual organization, and ultimately design. When Harvard established its Visual and Environmental Studies (VES) department in the 1960s, Loeb joined as an inaugural member and taught until 1989 on the graph-theoretic structures underlying polytopes, partitions, packings, and other visual structures.

Loeb saw crystallographic calculations as an indispensable tool of the emerging design scientist, a new kind of visual practitioner whose concerns could span architecture, engineering, and systems design. Loeb injected crystal methods into his curricula at Harvard's Graduate School of Design and in the Visual and Environmental Studies department. His courses, such as Introduction to Design Science, Synergetics: The Structure of Ordered Space, and Symmetry and Transformation, proposed to "develop inductively techniques and principles for dealing quantitatively with spatial complexities" that crossed the boundaries of disciplines.[40] Natural forces and geometry were two sides of the same coin for Loeb. He asserted that space itself was an active organizing medium: "Space is not a passive vacuum, but has properties that impose powerful constraints on any structure that inhabits it. These constraints are independent of specific interactive forces, hence geometrical in nature."[41] The techniques developed in Loeb's courses coincided significantly with Huff's methods, encompassing

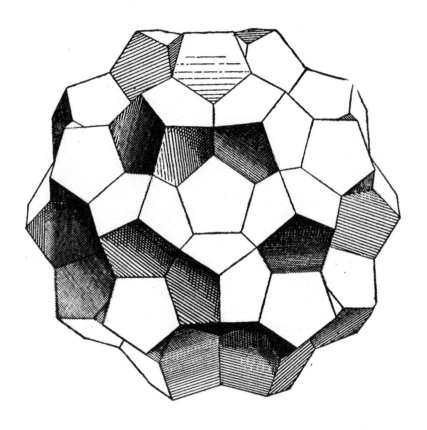

10.9   Drawing of a mathematical crystal model available for sale from a nineteenth-century
catalog of mathematical models.   Source: Martin Schilling, *Catalog mathematischer Modelle*
(Halle: Martin Schilling, 1903), 119.

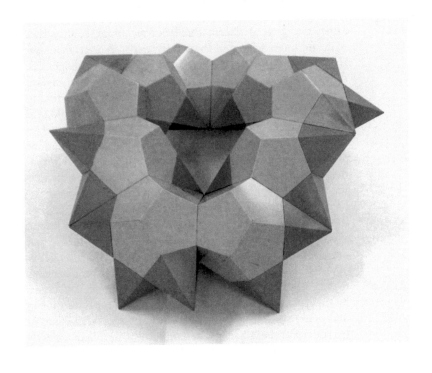

10.10 A student model from Ulm HfG, circa 1963.
Source: Ulm HfG Archive, photo by the author.

plane-filling tessellations; linear, homological transformations; symmetric solid dissections; and wallpaper and frieze pattern groups. Other themes included sphere packings and a range of tensegrity and dynamic structures topics. These courses became a conduit for applying rigorous methodologies of discrete mathematics and crystallography in design, often in very tangible ways: the organizational researcher Amy Edmondson, a student in several of Loeb's classes, went on to work as one of Buckminster Fuller's last engineers, and wrote the definitive book on Fuller's methods.[42]

Loeb viewed crystallography as the nucleus around which a disparate range of other mathematical techniques could gather to create a new science of shape. His text *Space Structures: Their Harmony and Counterpoint* (1976) recast complex crystallographic drawing methods as armatures for architectural analysis. He elaborated on Victor Schlegel's 1883 method for the flattening of crystals into warped map projections, akin to a topological flattening of a room's interior elevations in a developed surface drawing.[43] Loeb also introduced several early design applications of Dirichlet domains, now more commonly known as Voronoi diagrams, demonstrating how they could be applied for districting schools, hospitals, or other networks of civic resources.[44] In Loeb's hands, even arcane crystal methods found fresh relevance in the wholesale repurposing of spatial mathematics for design.

Beyond its capacity to produce and regulate cellular spatial organizations, crystallography was a model for the interaction of physical forces in taut equilibrium. Through it, geometry was material, and vice versa. Architect Peter Pearce connected lattice mathematics with methods of dynamic material optimization in a curriculum that bridged mathematics, physics, and assembly in his courses at the University of Southern California in the early 1970s.[45] He was preoccupied by methods that would negotiate between intrinsic and extrinsic constraints, especially those that resulted in minimum-energy conditions. Pearce described form as a "diagram of forces" driven by "the notion that design goals—in fields such as architectural, furniture, product, and industrial design—should be based on the need to fulfill performance objectives.... The result is a responsible approach to design that avoids arbitrary, superficial, and style-oriented solutions."[46]

Crystallography was not merely a means to draw dense new spatial complexes but also a kind of proxy physics in graphic form. Pearce theorized the crystal and minimal-surface form in his prescient text *Structure in Nature Is a Strategy for Design* (1978). In it, he developed systematic families of both periodic and aperiodic minimal surfaces, framed by crystalline lattices, which become the kernels of an expansive design syntax. His follow-up *Experiments in Form: A Foundation Course in Three-Dimensional Design* (1980) argued for a resolutely functional geometric pedagogy realized in multiresolution representation of minimal-surface forms.[47] The serialized approach led students through a sequence of progressively more complex exercises in spatial cartography and representational transformations. The same fundamental minimal geometry was rearticulated variously as lattices of hyperbolic patches, egg crate structures, triangular facetizations, cubic voxelizations, and minimal sphere packings, with each transformation proceeding from specific and deterministic performance objectives. Crystal form became a framework for intense optimization of performance.

Loeb's and Pearce's books joined a burgeoning library of architectural texts that codified crystal and polyhedral geometry as architectural techniques. This library marked the emergence of a cultural consensus on values, language, and representation in crystallographic design. The earliest of these texts were collections of visual precedents that served as sourcebooks of crystalline representation. Perhaps in homage to Weyl, William Huff compiled his own *Symmetry*, a multivolume series of short pamphlets that assembled instances of crystal forms across visual and scientific culture. Huff's research circulated in influential architectural journals, for instance appearing in *Oppositions* regularly between 1974 and 1977, alongside essays by theorists Kenneth Frampton, Manfredo Tafuri, and Rem Koolhaas.[48] The pamphlets themselves were collected by influential designers and theorists of the 1960s and 1970s, including Cedric Price of Archigram. Other authors developed textbooks of geometric technique applied to architectural problems of organization and ornamentation. Keith Critchlow's *Order in Space* (1969), which he touted as a "design sourcebook," offered a teaching manual of polyhedral form, with prolific illustrations showing the ornamental and organizational uses of such structures. Critchlow's avowedly visual approach, tuned to the designer's sensibilities, was intended to "allow the hand, eye, and mind to work together" in a new intuition of crystal form.[49] Many of the methods were highly algorithmic, and precisely presaged many of the fundamental operations of architectural computation.

As crystallography infiltrated academic design, a common set of mathematical practices and theoretical priorities were distilled into transmissible, teachable, and programmable knowledge systems. Design treatises influenced by mathematical texts became an essential medium to explain and perpetuate the intricate methods of polyhedral and crystal geometry to architects. Encoded in taxonomized manuals, these conventions set the stage for methods of crystal design to be readily transposed into digital formats for programmed computational processing.

Cellular Calculation, Visual Computation

The discrete mathematics of crystallography was ready-made for discrete computation, and the reticulated architectures of crystal mathematics reached new heights of intensity through the calculational power of digital computers. In fact, in their more complex manifestations, the cumbersome matrix operations required to actually calculate nonstandard lattice spaceframes made the computer almost indispensable.[50] Institutions incubated the early seeds of crystal computation, with Tomás Maldonado later recalling of Ulm "that the ideas that many of us had in the 1950s, that a kind of symbiosis of calculation and graphic representation could be created within the process of problem-solving, is basic to the widespread present-day use of computer graphics techniques."[51] In fact, computation was practically baked into the curriculum at Ulm. In 1967 William Huff wrote "The Computer and Programmed Design," a text which drew the long arc from the already antiquated methods of modernist design education toward a mathematical pedagogy of algorithmic form:

> The traditional foundation courses of the Bauhaus variety have proved in recent years to be arbitrary and lacking in objectivity for the purpose of design education. In certain

reformed first-year programs, a cohesive theory of structure has been developed; beginning design students are being introduced to studies in symmetry, topology, perception, and the raster phenomenon, upon all of which design exercises have been constructed. . . . The revolutionized basic design studies and the revolutionary computer graphics imply potentials of mutual benefit. . . . The computer can, without doubt, speed up the whole laborious investigatory process, demanded by thorough research, in exhausting all possible combinations and variations that any students' findings may suggest.[52]

The calculational capacities of the computer implied a qualitative rupture in how design could be conceived and taught. The deep structures of form, such as crystal lattices, became the essential content of design knowledge. Software fueled the fire of design research into proto-parametric architectures, as crystal patterns morphed into algorithmic logics.

In the 1970s, the transscalar character of crystalline modules allowed them to be multiplied as a structural system of almost infinitely variable size. With the computer, microcosm became macrocosm as molecular bonds hypertrophied into entire architectures. The most durable manifestation of computational crystallography in architecture was the explosion of polyhedral spaceframes of every description, at every scale. When this elemental structural rationale was paired with the computer, static Platonic configurations melted into liquid parametric structures. Skeletal frames, with limbs and joints of simulated steel, drew more explicitly biological connections, emphasizing that the animate and inanimate were being grafted together in a single calculated matrix. Tetrahedral or cuboctahedral tessellations danced in orchestrated unison, the bones and muscles of architectural organisms.

Speculative fantasies from buildings to structural systems were computational playgrounds for architects of the early 1970s. A consistent leitmotif was the molecular lattice as a mental model of spatial organization. On the schematic level, the German architect Ludwig Rase's (1925–2009) exhibition at the 1970 Venice Biennale showed his full embrace of the computational spaceframe as a

fig. 10.11

10.11   An example of Ludwig Rase's polytopic assemblages.
Source: "Ludwig Rase: Two- and Three-Dimensional Computer Design," *NOVUM Gebrauchsgraphik* (August 8, 1972): 53.

design parti. Moments in an endless cosmos of cuboctahedra, his structures flowed from a paradoxical but absolute faith in the power of modularity to sponsor flexibility and iterative evolution. Like Tyng, Rase saw the layout problem of designing a floor plan as analogous to concocting chemical compositions:

> The programming of each of these unit functions is done on the basis of an "atomic model." Each module is regarded as an atom which may fulfill a great variety of functions; with the aid of certain combination priorities the number of living-units can be combined in one single living molecule. The spatial view required in this sort of planning by far surpasses the designer's powers of mental visualization. With the aid of computers, however, the proper solutions can be visualized so that optimal groups of living-units result.[53]

Commercial ventures rushed to capitalize on the structural advantages of crystalline frames, developing some of the first computer numerically controlled (CNC) fabrication methods applied to architecture. Among other ventures, Peter Pearce founded Synestructics to leverage his crystallographic expertise in the development of modular nodal systems.[54] Pearce's forms were area-minimizing yet volume-maximizing, and like Rase's were designed for maximum flexibility and formal freedom. In this context, Pearce developed his novel "nodeless" spaceframes, driven by his own bespoke CNC-controlled welding software.[55] Compared to the better-known Buckminster Fuller, Pearce felt no ideological allegiance to the sphere as ur-form. Instead, his structures afforded much broader versatility in terms of the overall design of building shape. The most famous use of these frame structures was as the basis for his design of Biosphere 2, a hermetically enclosed experiment in simulated ecology, which could be seen as a metaphor for the increasingly hermetic world of crystallographic architecture itself. As crystal research matured, it drove both conceptual approaches to design and entirely new business models for bespoke structures.

The Crystallization of a Thought Collective

Through innervated webs of personal correspondence, disciplinary conferences, research institutes, and sympathetic journals, bundled threads converged over the course of the 1970s into a coherent subculture of crystallographic architecture. Ecumenical by nature, it drew architects, mathematicians, and other scientists to develop a shared machinery of lattice form. Arthur Loeb and Peter Pearce were central protagonists in the thickening web of social and research connections that nourished the heart of this new subculture. Loeb gathered around himself a sizable network of interlocutors, both designers and scientists, which collectively spanned every conceivable aspect of what he termed "design science." Many of these eclectic connections were academic interlocutors, including Buckminster Fuller, whom Loeb had met at the University of Pennsylvania in 1973. The two would review each other's work, and Loeb would often assist in Fuller's highly theatrical lectures. Loeb also made significant contributions to Fuller's *Synergetics*, including writing the preface to the book, and introduced some of Fuller's ideas to his own extensive body of students. Beyond his connection to Fuller, Loeb also carried on an extensive correspondence and developed a warm friendship with his Dutch compatriot M. C. Escher (1898–1972), whose half-brother Berend George Escher (1885–1967) was a crystallographer and volcanologist whom Loeb knew from academic conferences. Unsurprisingly, Escher was by far the most referenced visual artist in Loeb's classes, and deformations similar to Huff's parquet transforms appeared frequently in Loeb's courses as well.

fig. 10.12 Loeb organized the social project of crystallographic design along scientific lines, conforming architectural conversations to the conventions of conferences and publications in the hard sciences. He convoked his diverse interlocutors in a long-running series of design science conferences, and edited Birkhauser's book series *The Design Science Collection*, which became an intellectual home for the maturing crystal movement. He also mentored young architects like Pearce, who in turn attended many of the design science symposia that Loeb organized. Their relationship was close enough that Pearce specially acknowledged Loeb in his

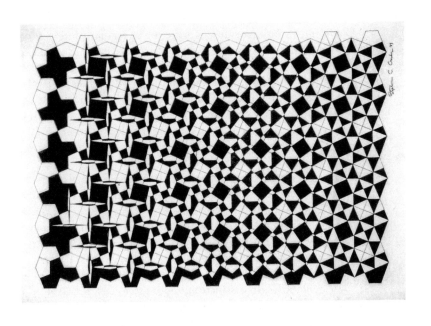

10.12　Student exercise for a parquet deformation, ves 175: Introduction to Design Science.
The drawing was computer-programmed and automatically generated. Instructor: Arthur
Loeb. Student: Cohen, Fall 1981, Box 11 (Slide box 2.33).　　Source: Harvard University Archives,
Carpenter Center Teaching Collection, uav 869.5295.1

Structure in Nature Is a Strategy for Design. Like Loeb, Pearce significantly contributed to Fuller's *Synergetics*: for nearly a year and a half, Pearce assisted Fuller in the preparation of the book, including drafting the book's detailed diagrams.[56]

Pearce was an interdisciplinary connector in his own right, refining his intuitive sense of structure by regularly interacting with mathematicians of both crystal forms and minimal surfaces. Throughout *Structure in Nature*, he referenced and built on the work of Alan Schoen, the mathematician who pioneered triply periodic minimal surfaces and who was in conversation with other designers of the time, including the architectural sculptor Erwin Hauer. Schoen, in turn, referenced Pearce's work in his seminal 1970 NASA report *Infinite Periodic Minimal Surfaces without Self-Intersections*.[57]

This loose collection of associations and informal personal collaborations between mathematicians and designers gradually gave way to more formalized cross-disciplinary research clusters that signaled a maturing thought collective. Foremost among them was the University of Montreal's Groupe de Recherche Topologie Structurale, established in 1978 as a cluster dedicated to graph-theoretic and visual investigation of faceted polytopic geometries. In distinction to pure topology, the Groupe's structural topology integrated dimensional geometry with topological organization. Founded jointly under the auspices of the university's Center for Research in Applied Mathematics and the Department of Architecture, the Groupe aimed to marshal techniques of both design and mathematics into a unified science of formal systems. The Groupe's advisory board, composed of architects and mathematicians, included the architects Moshe Safdie (who claimed he nearly became a mathematician himself)[58] and Lionel March. Among the membership was Steve Baer, a designer originally trained in mathematics at ETH, and author of the influential *Dome Cookbook* (1967). The Groupe de Recherche Topologie Structurale realized an oft-dreamed fantasy of the truly cross-disciplinary collective whose aim was the rigorous synthesis of design and mathematical culture.

The Groupe disseminated its research through the journal *Topologie Structurale*, addressed to "structural engineers, architects, and mathematicians who are interested in the fundamental

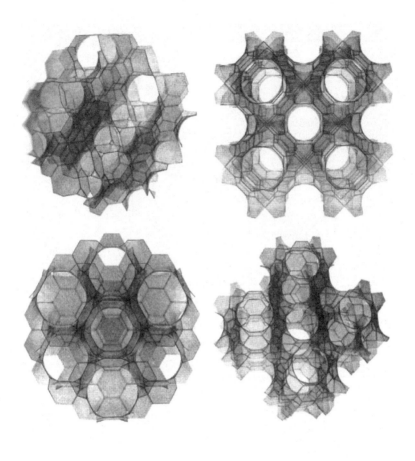

10.13   Peter Pearce's triply periodic minimal surfaces.
Source: Peter Pearce, *Structure in Nature Is a Strategy for Design* (Cambridge, MA: MIT Press, 1978).

10.14   Alan Schoen's minimal-surface models.
Source: Alan H. Schoen, *Infinite Periodic Minimal Surfaces without Self-Intersections*
(Washington, DC: NASA, 1970).

problems concerning three-dimensional space."[59] An apotheosis of increasingly symbiotic design science research, the journal was a comprehensive fusion in which diction, symbology, and descriptions of a thoroughly mathematical character were married to drawing methods akin to architectural representations. Published from 1979 to 1997, *Topologie Structurale* represented an apex of the crystallographic thought collective—a unified forum for voices from diverse disciplines tackling common problems in visual mathematics.

The supremely precise language of *Topologie Structurale* was a mixed virtue, a signal of the maturation of this architecture-mathematics axis but also evidence of the creative exhaustion of crystal discourse. The technical lexicon that described complex form demanded a deep initiation into underlying mathematics, and it ultimately frustrated the Groupe's interest in broadening an architectural conversation. If anything, the rigorously scientific frame for architecture was a pyrrhic victory, achieved at the cost of a more direct engagement with contemporary design culture, by then in the throes of postmodernism. As a result, a thought collective of unique utility turned insular and marginal precisely when its geometric rigor could have shaped the emerging culture of digital design.

Computable Crystals: A Hypercubic Sociology

One of the most incisive distinctions in Ludwik Fleck's theory of the "thought collective" is between esoteric and exoteric circles within a subculture: "there arises, round every product of collective life, a smaller esoteric circle, composed of members having a more direct relationship to this product, and a larger exoteric circle composed of members who participate in it through the intermediation of the other ones."[60] Despite the intensely rigorous development of an esoteric crystal thought collective, there were countervailing architecturalizations of lattice forms, exoteric experiments in which breaks, fractures, and parallel courses in crystallographic design emerged. The crystal proved itself resilient and polysemous. American architect Peter Eisenman's 1988 project for the Carnegie Mellon Research Institute (CMRI) was characteristic of such critical tendencies.

Ignoring the more orthodox approach to lattice form exemplified by *Topologie Structurale*, Eisenman proposed a hypercubic crystal understood not only as a structural and spatial frame, but also as a social and communicative organization to accelerate the creation of knowledge.[61]

The catalyzing crisis of the Carnegie Mellon project was the problem of knowledge creation in the digital age. The institute was to house a highly interdisciplinary set of researchers, combining laboratories for material and computer science with biochemical facilities and artists' studios. The aspiration was to develop an environment of truly serendipitous collaboration and a design which intensified, as much as possible, the social and research interactions between these disciplines. Carnegie Mellon's president, Richard Cyert, set the project's brief:

> The problem today for man is to overcome knowledge: you see, computers have knowledge, robots have knowledge.... The knowledge revolution, artificial intelligence, and the systems of knowledge ... have gotten out of hand. ... Science today is trying to find a way to control knowledge, and the knowledge revolution.... I want you to do a building which symbolizes man's capacity to overcome knowledge.[62]

Eisenman hoped to transpose hyperdimensional crystal structures used in artificial intelligence (AI) to short-circuit the human communication environment and, ipso facto, the collaborative creation of knowledge. He saw the radical potential of American computer scientist Daniel Hillis's "connection machine"—a massively parallel supercomputer for artificial intelligence and image-processing applications—to provide a model for intensified human interactions.[63] An annotated copy of Hillis's AI research found in Eisenman's project archive highlights the geometric dynamo of the project: the hyperdimensional n-cube. The n-cube's novel organization folded sheets of thousands of processors upon each other in superimposed matrices. In this configuration, the path distances and latencies between any two processors were dramatically reduced and the number of connections exponentially increased, allowing the whole collection of processors to function as a unified ganglial complex. Landscape was folded into volume, following the same morphological principle as the folding of the brain's cortex into a neuronal bundle. The social serendipity that drove research collaboration could be engineered through synaptic densification, and the result would be a cybernetic and autonomic innovation engine.

The peculiar properties of the n-cube, or Boolean cube, were inextricably tied to its crystal symmetry and its vectorially defined connections, as Eisenman recounts:

> The multiplication of N-geometries allows multiple paths for information movement so that, for instance, from any point in a 1000-N cube a move can be made in 1000 different directions within the information matrix.... Thus the Boolean cube is a complex structure which lies between the purity of a platonic form and the infinite and unlimited form of a non-Euclidean structure.... Each pair contains two solid cubes and two frame cubes corresponding to office and laboratory modules. Each pair can be seen as containing the inverse of each other in solid and void.[64]

Once this n-cube was defined, the other design moves followed serially. Pairs of n-cubes began to overlap and intersect, creating intricate superpositions and intersections reminiscent of gestalt overlays or anaglyphic stereograms. Finally, these pairs of cubes rotated, permuted, and transformed in a linear sequence in an infinitely extensible chain or intricately interlacing braids.

The parallels with crystal growth were not lost on Eisenman. Also in the project archives is a copy of the remarkable poem "Fluorite" by Roald Hoffmann (b. 1937), a Nobel Prize–winning chemist and erstwhile poet:

> "Minerals in their matrix
> are what I like best."
> Fluorite wears a variable habit.
>
> ...

A specimen I have
tumbles in inch-long cubes,
superimposed, interpenetrating,
etched on all their faces.

…

Eerie crystal.
Were a Martian photograph
enlarged to reveal such polyhedral
regularity, it would be deemed
a sign of intelligence at work. But
the only work here, and it is free
is that of entropy.[65]

In these curious verses there are clues to Eisenman's alternate crystal formulation: tumbling, imperfect, entropic forms with an echo of regularity but not beholden to the strict perfection of the crystallographic matrix.

Unfortunately, by late 1989, budget considerations and the university president's increasing ambivalence ensured that the Carnegie Mellon project would not move forward with Eisenman. In its place, the project has a mirror twin of its own designed by the firm Bohlin Cywinski Jackson, complete with open-frame cubes—which, unfortunately, are only three-dimensional.

In the Carnegie Mellon project, we see the reinterpretation of a crystalline electronic diagram as a format for human communication. While the project's geometric genesis—a complex chain of interpenetrating hypercubes—recalls recombinant dissections and parquet deformations, Eisenman's conceptual interests were not only formal. Crystal methods and motifs—like anaglyphic lattices recast as architectural drawings—constructed a sociology that mirrored computation but departed from the strictly geometric investigations like *Topologie Structurale*. The project embraced new and broader associations of crystallographic design, expanding its exoteric territory in fresh and vital directions.

Inanimate and Reanimated

Of the many varieties of crystal symmetry, mirrored reflection is perhaps the most intuitive and fundamental. Reflection cleaves reality in two: the actual and the virtual, original and duplicate, chiral and antichiral, object and representation. Yet as a generative technique, reflection also differs profoundly from more conventional architectural techniques of projection. In *The Projective Cast*, Robin Evans argued for the primary, if equivocal, role of geometric projection in architecture, both as historiographic lens and as a form-making armature.[66] In contrast, the optically reflected image was a dazzling but far more suspect type of representation: "Reflection, luster, refraction, luminosity, darkness, color, softness, absorption, liquidity, atmospheric density, instability of shape: these and a host of other properties jeopardize perceptions of metric uniformity."[67] Yet what Evans ignored was that mirror reflection and other nominally optical crystallographic operations promise a kind of space quite different from the conventions of projection. In multiplying reflections indefinitely across cells and aggregations, new structures transcend simple symmetries and projections. In speaking of the capacities of crystal structure, the polyhedral designer Steve Baer opined: "These are instructions on how to almost break out of prison. The prison is the paucity of shapes to which we have in the past confined ourselves because of our technology-industry-education-economy. . . . In the future the architect will work with volumes and areas with the same freedom that he has while playing with a balloon filled with x cubic inches of water."[68] Crystal structures unveiled a path toward a more liberated design, beyond static and limited notions of symmetry.

The evolution of specialized knowledge cultures within architecture challenges univocal disciplinary roles and boundaries, and the case of the crystallographic thought collective offers a clear example of the arc of one such knowledge culture. Within the cosmos of encounters between design and crystallography, diverse projects and architects reveal distinct clusters of knowledge acquisition and transmission. In particular, we see a vibrant thought collective flourishing around figures like Arthur Loeb, Peter Pearce, and William Huff, who constructed a common methodological machinery enabled by crystal symmetries. Influenced by both disciplinary and cultural fascinations, the designers in this cohort researched—and in turn, communalized—progressively more rigorous mathematical methods. Common frameworks extended to a relatively recognizable graphic language and set of drawing conventions. Clear and consistent pedagogies, sponsored indirectly by a Cold War penchant for positivist curricula, propagated this knowledge to a broader cohort of students. Around the technical sympathies of this nucleus coalesced more formal and robust media of exchange, such as Arthur Loeb's design science symposia or journals such as *Topologie Structurale*.

Which brings us full circle to Fleck's sociology. Among the members of the esoteric kernel of a subculture, words like "crystal," "geometry," and even "mathematics" become freighted terms of art, evoking nuanced and precise rules of virtuosic manipulation. But design cultures are rarely unitary, and at the exoteric periphery there are turbulent currents and countercurrents as other designers interpret scientific fascinations independent of or even contrary to the accepted conventions of the esoteric core. In this exoteric range are Eisenman and van Doesburg, each inventing an independent creative approach unbeholden to the mathematically self-conscious orthodoxy. Notably, neither of these exoteric designers aligned himself with an explicit *institutional* crystallographic research agenda. Instead, by choice or necessity, they both constructed their own alternative alibis for the use of the crystal in design. Even as the esoteric collective withered, this exoteric work took on new and separate life. The genesis and decline of an esoteric thought collective, as well as the collateral activities of its exoteric circles, are thus useful lenses through which to register the social dynamics of evolving design culture.

Ultimately, both disciplinary and exogenous factors conspired to kill crystallographic architecture during the 1980s. Internal exhaustion played a role: the esoteric crystal thought collective failed to transmit its considerable technical machinery to a new generation. The critical culture of architecture also did not value it, or failed to consume it in a progressive way. Cycles of architectural fashion took their toll. As spaceframes and their associated geometries became passé, the commercial motivation for lattice research in design grew less compelling. The mania for New Math faded, and was then discredited. Design Methods and

Design Research approaches, with which crystal tendencies were inextricably aligned, failed to deliver on exaggerated promises. Ironically, the general dissipation of crystallographic design occurred at the precise moment when such techniques would have been most powerful: when the cost of computation dropped precipitously with the advent of the personal computer, and the subsequent flourishing of digital design techniques.

An entire design subculture evaporated with the death of crystallographic architecture. Absent that rigor, digital technologies were theorized not through the systematic and intrinsic lens of procedural logic, but instead often through visual and literary analogy. This syncretic rhetorical approach tended to obscure or divert rather than enable the advancement of mathematical methods underlying digitization. There were exceptions to be sure, and the dynamism of early crystal investigations had parallels in this new discourse, notably through architect Greg Lynn's text *Animate Form*, which also engaged mathematical methods such as calculus in a substantial way.[69] The mathematical facts of crystal form were stubborn; even if they were neglected, they merely lay dormant, in suspended animation. Epistemic center and periphery were in constant flux, and as certain technical cultures died, others took up their source material with fervor. Perhaps it is precisely this cyclic death and reanimation which is the greatest testament to the vitality of subcultures in architecture and to the enduring relevance of crystal form.

11   Of Dabblers and Virtuosos

For the architect, the uniquely expansive symbolic language of mathematics encodes polarities of geometry and perception, science and aesthetics, culture and nature, bringing intellectual dichotomies into intimate creative contact. This act of encoding transcribes processes and rules from human knowledge into symbolic, mechanical, or electronic instruments that simplify and automate complex systems. Yet the technical initiation and meticulous training required for fluency in this very peculiar language has always proved a formidable barrier to its complete embrace in the discipline of architecture. Today, computational tools that encapsulate mathematical methods are radically short-circuiting the path to expertise, blurring and shading the facile distinction between dabbler and virtuoso, and democratizing access to the systems and aesthetics of mathematical design. Old hierarchies of training and knowledge are eroding, and in their place a new kind of technical culture of opportunistic hacking, open code sharing, and recombinant invention flourishes.

Expertise is an inescapable enabler of mathematical design, encompassing a spectrum between the poles of dabbler and virtuoso. The film critic A. O. Scott describes the dabbler as "caught in an awkward spot between the generative passion of creation, which drives and defines the artist, and the casual promiscuity of the consumer, who picks up and discards works of art or cultural experiences based on momentary whims or interests."[1] The assessment of a designer's place in this spectrum is subject to perspective: no less an architect than Christopher Wren, whose rectification of the cycloid curve assured his international mathematical fame, was judged by later historians of mathematics as a "skilled dabbler, an amateur."[2] Architectural design labors under an ambivalent relationship to scientific expertise. While architecture suffers no shortage of dabblers, the scientific dilettante has also long had a suspicious or even dangerous reputation as a charlatan. Max Bill, who was one of the earliest to import topological ideas into design, was nevertheless deeply dubious of the capacity of the discipline to absorb more sophisticated technical methods. He was particularly skeptical of the work of his successor at Ulm, Tomás Maldonado, and the increasingly scientistic program of the school: "of course it is [Tomás] Maldonado who is the worst influence—with his ambitions to introduce the loftiest scientific methods he opens the floodgates of wretched dilettantism."[3] The anxieties of some architects concerning the ascendant role of mathematical knowledge in design presaged a broader disciplinary angst surrounding the computer and the mathematical initiation its use presumed. The essayist of a 1971 piece in *Progressive Architecture* offered that "the new math has become so standard that 'Base 5' and 'Base 2' are just a couple more things that you have to learn before recess. But there are also those in the design professions to whom 'Base 2' is still the symbolic stumbling block to taking advantage of the computer as a valuable office and professional tool."[4] Surprisingly, Antoine Pevsner was yet more pointed in his criticism of dabblers:

> At the time of Cubism and Futurism, many, inspired by geometry, chose to represent objects through an acquaintance with descriptive and analytic geometry and certain physical phenomena in an arrangement of geometric forms in a kind of scientific or mechanical project; they believed that artists should employ logic rather than intuition in their work. Sometimes their constructed pictures were striking, but like soap bubbles, they were exciting for only a little while.[5]

Even those who might normally be seen as the avatars of scientific design were reflexively cautious of its callow misuse.

At the polemic antipode to the dilettante's naiveté is the virtuoso's hard-won expertise. The virtuoso has, by dint of will, exhaustively probed a particular territory of mathematical form—warped surfaces, voxelized combinatorics, topological contortions—and achieved fluent facility in its arcane logic. Such virtuosity propels many of the most elevated expressions of exact design, pushing back the frontiers of geometric and spatial possibilities. Yet the devoted labor necessary to achieve this expertise drastically limits its diffusion. This exclusivity is compounded by the social frameworks that ossify around any technically specific subculture. The very structures that nurture expertise—term-of-art diction, esoteric thought collectives, subdisciplinary institutions, thematic laboratories—tend to cloister promising research from the broader culture of design. The common impulse to hothouse expertise through institutions which support but ultimately segregate that expertise—like Cambridge's Centre for Land Use and Built Form Studies, or Montreal's Groupe de Recherche Topologie Structurale—creates archipelagos of distinct but hermetic technical subcultures. This tends to nurture nascent subcultures to develop into more autonomous subdisciplines, at the high price that their practitioners may be increasingly alienated from the broader critical dialog of the mother discipline of design.

Digital methods of black-boxing technique in software are quietly remixing fine distinctions between dilettante and virtuoso. The alchemy of software makes the products of expertise attainable by the novice, and the expertise itself is rendered nearly superfluous. Meticulous training, careful attention, and conceptual fluency are no longer as essential to experimentation with mathematical forms. Institutionalized social subcultures, once the primary incubator for methodological innovation, are displaced by ad hoc networks of software distribution or on-demand tutorials available to individuals. As a repository for encapsulated expertise, digital encodings subvert labored training and cultural initiation and replace it with an opportunistic rewiring of patchwork techniques and workflows. The computer thus dissolves accepted hierarchies of knowledge in favor of a promiscuous and polyglot patchwork of representation and technique, accessible to the uninitiated layman. The divorce of expertise from technique extends beyond processes of creation to modes of perception, as visual intuition itself increasingly convolved with black-boxed computational filters like automatic image classifications, machine vision protocols, deep learning, and artificial intelligence. What is happening is not a deskilling of architecture and perception in the Marxist sense but quite the opposite. Various alternative modes of geometrical and sensorial training mold increasingly hybridized spatial, graphic, and visual intuitions shared among geographically distributed cohorts. Software has created an accessible, rich, expansive atlas of technique, a profound reformatting of how knowledge is consumed and deployed. We are all dabblers now.

Yet, despite the frictionless consumption of mathematical technique underwritten by software, the coherence of technical subculture remains as indispensable as ever for forging loose aesthetic affinities in common intellectual projects. What has changed decisively are the contours of these collectives: they may be smaller or larger, more diffuse or more focused, more geographically local or more distributed than in the past. Specific ways of seeing and making have become regularized and regimented through code, but these codes themselves become merely building blocks for vastly differing recombinant practices of design. Collectives today are not only social but also technological, absorbing the logic of distributed information systems directly into the culture of the collective itself.

What, then, is the place of mathematical expertise in an architectural discipline surfeited with ready-made technique distilled into consumable software components? A culture veined with hidden coded processes demands critics of that medium, and mathematics becomes a Rosetta stone for decoding, interrogating, and hacking not only digital technique but culture itself. In this world, the scientifically inclined designer is uniquely equipped as a critic of calculational regimes and their cultural implications. Her inside-out knowledge, honed by creative application, opens avenues of judgment unavailable with exclusively historical or sociological analysis. She recognizes and attends to the details of calculational technique in the way a painter would recognize the mixture of a tempera. More fundamentally, to bend mathematics toward design is itself a critical act, an act of radicalism against the complacencies of taste and style alone. It is a claim that creative and cultural work must be tried in the crucible of logic. It is a resolute affirmation that knowledge, technique, and architecture are not naturally segregated but must be subject to the most rigorous reflection at their intersection in creative production.

Mathematical quantification articulates and harmonizes the disparate demands and aspirations of architecture, a compromised discipline constrained by myriad externalities that force it to operate along "the line where art and utility meet."[6] Mathematics is a crucible for the friction between what must be and what could be. The lingua franca of mathematical encoding is an interpretive medium that precisely brings incommensurate polarities into mutual conversation: art and utility, technology and nature, sensation and logic, design and science. In his essay "Man the Technician," the philosopher of history José Ortega y Gasset mused that humanity is "made of strange stuff as to be partly akin to nature and partly not, at once natural and extra-natural, a kind of ontological centaur, half immersed in nature, half transcending it."[7] In design, too, the most potent acts of imagination are those that vivisect the rigor of fact with the transcendent dreams of possibility. Mathematics sits at the nexus of these two, sometimes as a proxy for nature, sometimes as an engine of visionary new fantasies. At that nexus it offers evolving augmentations of design practice. It furnishes exact formats for inexact natural forms, marshals hyperdimensional encodings of multifarious design objectives, opens a topological gradient between once-distinct spatial categories like interior and exterior, constructs new perceptual media for the stereoscopic and photographic projection of exact imagination, or acts as an exact calculational proxy for the natural sciences of physics, biology, or chemistry. Across its polyvalent roles, mathematics can be a hermeneutic lens for interpreting (or artfully misinterpreting) nature, an armature for prosthetic vision, and an accelerant and exact language for the creative acts of architecture itself. Mathematics transcribes architecture from the drawing into a different kind of text, amenable to peculiar new modes of interpretation. What matters is not so much the fidelity of the translation but its intellectual and aesthetic productivity—and by that measure, the studies in this book attest to the fecundity of mathematical design as a creative practice. Of course, there is a grain of unattainable fiction in this dream. As the Brazilian author Clarice Lispector enthused, "I don't want the terrible limitation of those who live merely from what can make sense. Not I: I want an invented truth."[8] Architecture through the lens of mathematics is a kind of invented truth, a language of both necessity and imagination, a territory of elemental invention—rigorous, designed, yet unbounded.

Notes

Chapter 1

1 Robert Heinlein, "And He Built a Crooked House," *Astounding Science Fiction* 26, no. 6 (February 1941): 69.

2 Max Bill, "The Mathematical Approach in Contemporary Art," *Theories and Documents of Contemporary Art: A Sourcebook of Artists' Writings*, ed. Kristine Stiles and Peter Selz (Berkeley: University of California Press, 1996), 75.

3 Brian Rotman, "Thinking Dia-Grams: Mathematics and Writing," in *The Science Studies Reader*, ed. Mario Biagioli (New York: Routledge, 1999), 433.

4 Le Corbusier, *The Modulor* (London: Faber and Faber, 1954), 71.

5 Colin Rowe, *The Mathematics of the Ideal Villa and Other Essays* (Cambridge, MA: MIT Press, 1976), 9.

6 José Ortega y Gasset, "Man the Technician," in *History as a System* (New York: W. W. Norton, 1941), 95.

7 Robin Evans, *The Projective Cast* (1995; repr., Cambridge, MA: MIT Press, 2000), 348.

8 Jonathan Crary, *Techniques of the Observer* (Cambridge, MA: MIT Press, 1990), 2.

9 Soraya de Chadarevian, *Designs for Life: Molecular Biology after World War II* (Cambridge: Cambridge University Press, 2011), 137.

10 Bruno Latour, *Science in Action: How to Follow Scientists and Engineers through Society* (Cambridge, MA: Harvard University Press, 1987), 2–3.

11 Johanna Malt, *Obscure Objects of Desire: Surrealism, Fetishism, and Politics* (Oxford: Oxford University Press, 2004), 167

12 John Elsner and Roger Cardinal, introduction to *The Cultures of Collecting*, ed. John Elsner and Roger Cardinal (London: Reaktion, 1994), 2.

13 Ann Blair, *Too Much to Know: Managing Scholarly Information before the Modern Age* (New Haven: Yale University Press, 2010), 2.

14 Peter Galison, "Judgement against Objectivity," in *Picturing Science, Producing Art*, ed. Caroline Jones and Peter Galison (New York: Routledge, 1998), 328.

15 Galison, "Judgement against Objectivity," 329.

16 Massimo Scolari, *Oblique Drawing: A History of Anti-Perspective* (Cambridge, MA: MIT Press, 2015), 151.

17 Ludwik Fleck, "The Problem of Epistemology," in *Cognition and Fact: Materials on Ludwig Fleck*, ed. Robert S. Cohen and Thomas Schnelle (Dordrecht: Springer, 1986), 79–112.

18 Lorraine Daston, ed., *Biographies of Scientific Objects* (Chicago: University of Chicago Press, 2000), 13.

19 Quoted in Michelle Kuo, "'Inevitable Fusing of Specializations': Experiments in Art and Technology," in *Rauschenberg*, ed. Leah Dickerman (New York: Museum of Modern Art, 2016).

20 Heinlein, "And He Built a Crooked House," 69–70.

21 Heinlein, "And He Built a Crooked House," 75.

22 Heinlein, "And He Built a Crooked House," 81.

Chapter 2

1 Stéphane Van Damme, *Paris, capitale philosophique: De la Fronde à la Révolution* (Paris: Odile Jacob, 2005), 147.

2 Lorraine Daston, "Nature by Design" in *Picturing Science, Producing Art*, ed. Caroline Jones and Peter Galison (New York: Routledge, 1998), 239.

3 C. R. Hill, "The Cabinet of Bonnier de la Mosson (1702–1744)," *Annals of Science* 43 (1986): 152.

4 Hill, "The Cabinet of Bonnier de la Mosson," 151.

5 Karin Knorr Cetina, *Epistemic Cultures: How the Sciences Make Knowledge* (Cambridge, MA: Harvard University Press, 1999), 1.

6 Y. Kalay, L. Swerdloff, B. Kajkowski, and C. Neurmberger, "Process and Knowledge in Design Computation," *Journal of Architectural Education* 43, no. 2 (2014): 50.

7 My use of the term "instrumental knowledge" is distinct from its typical sense in the epistemology of science, where it refers to theories of predictive reliability (and thus instrumentality), as in Richard Boyd, "Scientific Realism and Naturalistic Epistemology," *Symposia and Invited Papers*, vol. 2 of *Proceedings of the Biennial Meeting of the Philosophy of Science Association* (1980): 614.

8 Benjamin Aranda and Chris Lasch, *Tooling* (New York: Princeton Architectural Press. 2006), xi.

9 Richard N. Langlois, "Knowledge, Consumption, and Endogenous Growth," *The Journal of Evolutionary Economics* 11, no. 1 (2001): 85.

10 Frank D. Prager and Gustina Scaglia, *Brunelleschi: Studies of His Technology and Inventions* (London: Dover, 1970), 135.

11 Leon Battista Alberti, *On the Art of Building in Ten Books*, trans. and ed. Joseph Rykwert with Neil Leach and Robert Tavernor (Cambridge, MA: MIT Press, 1988), 3.

12 Alberti, *On the Art of Building*, 3. The original Latin passage reads, in part, "Fabri enim manus architcto pro instrumento est." Leon Battista Alberti, *De re aedificatoria* (Strasbourg: Jacobus Cammer, 1541), 1.

13 Massimo Scolari, *Oblique Drawing: A History of Anti-Perspective* (Cambridge, MA: MIT Press, 2015), 250.

14 Mario Carpo and Frédérique Lemerle, eds., *Perspective, Projections and Design: Technologies of Architectural Representation* (London: Routledge, 2008), 50.

15 Prager and Scaglia, *Brunelleschi*, 105.

16 Robin Evans, *The Projective Cast: Architecture and Its Three Geometries* (1995; repr., Cambridge, MA: MIT Press, 2000), 187.

17 Philippe Potié, *Philibert de L'Orme: figures de la pensée constructive* (Paris: Editions Parenthèses, 1996), 9.

18 E. M. Horsburgh, *Modern Instruments and Methods of Calculation* (London: G. Bell and Sons, 1914), 256.

19 George Adams, *Geometrical and Graphical Essays Containing a General Description of the Mathematical Instruments Used in Geometry, Civil and Military Surveying, Levelling, and Perspective* (London: J. Dillon and Co., 1797), 152. George Adams's *Geometrical and Graphical Essays* (1791) first introduced Suardi's work to an English-speaking audience. Adams's work was first published in German in 1795.

20 Adams, *Geometrical and Graphical Essays*, 151.

21 Gerard L'Estrange Turner and Margaret Weston, *Nineteenth-Century Scientific Instruments* (Berkeley: University of California Press, 1983), 277.

22 William Ford Stanley, *A Descriptive Treatise on Mathematical Drawing Instruments* (London: Butler and Tanner, 1878), 77.

23 Francis Cranmer Penrose, *The Principles of Athenian Architecture, or The Results of a Survey Conducted with Reference to the Optical Refinements Exhibited in the Construction of the Ancient Buildings at Athens* (London: Macmillan, 1888).

24 Stanley, *Descriptive Treatise*, 85.

25 Cyril M. Harris, *American Architecture: An Illustrated Encyclopedia* (New York: W. W. Norton, 2003), 164.

26 *Journal of the Franklin Institute* 2, no. 2 (August 1826): 106–108.

27 F. N. Massa, "An Instrument for Describing Mathematical Curves," *Scientific American* 89, no. 7 (August 15, 1903): 116.

28 Robert J. Whitaker, "Harmonographs. II. Circular design." *American Journal of Physics* 69, no. 2 (2001): 175.

29 Andrew J. Butrica, "The Mind's Eye: Technical Education, Drawing and Meritocracy in France, 1800–1850," *Icon* 21 (2015): 9.

30 Gaspard Monge, *Géométrie descriptive* (Paris: Baudouin, 1799; repr., Paris: Éditions Jacques Gabay, 1989), 4.

31 Jean Paul Douliot, *Traité spécial de coupe des pierres* (Paris: Imprimerie de Richomme, 1825), 59.

32 Charles-François-Antoine Leroy, *Traité de stéréotomie, comprenant les applications de la géométrie descriptive* (Paris: Bachelier, 1844), 64.

33 Louis Mazerolle, *Traité théorique et pratique de charpente* (Paris: Editions H. Vial, 1895).

34 Douliot, *Traité spécial de coupe des pierres*, 149.

35 Lorraine Daston, "Enlightenment Calculations," *Critical Inquiry* 21, no. 1 (Autumn 1994): 186.

36 Alison Morrison-Low, *Making Scientific Instruments in the Industrial Revolution* (Aldershot, UK: Ashgate, 2007), 188.

37 Morrison-Low, *Making Scientific Instruments*, 187.

38 Morrison-Low, *Making Scientific Instruments*, 188.

39 Stanley, *Descriptive Treatise*, 249.

40 Daniel Alexander Murray, *An Elementary Course in the Integral Calculus* (American Book Company, 1898), 188.

41 James Clerk Maxwell, "Description of a New Form of Planimeter, an Instrument for Measuring the Areas of Plain figures Drawn on Paper," *Transactions of the Scottish Society of Arts* 4, part 4 (1855).

42 Stanley, *Descriptive Treatise*, 68.

43 Arthur Ashpitel, *Treatise on Architecture, including the Arts of Construction, Building, Stone Masonry, Arch, Carpentry, Roof, Joinery, and Strength of Materials* (Edinburgh: Adam and Charles Black, 1867), 183.

44 Mario Carpo, "Drawing with Numbers: Geometry and Numeracy in Early Modern Architectural Design," *Journal of the Society of Architectural Historians* 62, no. 4 (December 2003): 448.

45 John A. N. Lee, *International Biographical Dictionary of Computer Pioneers* (New York: Taylor & Francis, 1995), 171.

46 Maya Hambly, *Drawing Instruments, 1580–1980* (London: Sotheby's Publications, 1988), 91.

47 Lee, *International Biographical Dictionary*, 173.

48 David Mills, *Difficult Folk? A Political History of Social Anthropology* (Oxford: Berghahn Books), 15.

49 Le Corbusier, *Towards a New Architecture*, 227.

50 Antoine Picon, "Architecture and Mathematics: Between Hubris and Restraint," *Architectural Design* 81, no. 4 (2011): 35.

51 Alfred North Whitehead, *An Introduction to Mathematics* (New York: Henry Holt, 1911), 61.

Chapter 3

1 Davis Baird, *Thing Knowledge: A Philosophy of Scientific Instruments* (Berkeley: University of California Press, 2004), 24.

2 Gaspard Monge, *Géométrie descriptive* (Paris: Baudouin, 1799; repr., Paris: Éditions Jacques Gabay, 1989), 4.

3 Robert Brain, *The Pulse of Modernism: Physiological Aesthetics in Fin-de-Siècle Europe* (Seattle: University of Washington Press, 2015), 12.

4 Leonhard Euler, "De saidis quorum superficiem in planum explicare vicet," *Novi commentarii academiae scientiarum imperiatis Petropolitanae* 16 (1772): 3–34.

5 Florian Cajori, "Generalizations in Geometry as Seen in the History of Developable Surfaces," *American Mathematical Monthly* 36, no. 8 (October 1929): 434.

6 Amy Shell-Gellasch and Bill Acheson, "Geometric String Models of Descriptive Geometry," in *Hands On History: A Resource for Teaching Mathematics*, ed. Amy Shell-Gellasch (Washington, DC: Mathematical Association of America, 2007), 49–62.

7 William Mueller, "Mathematical Wunderkammern," *Mathematical Association of America Monthly* 108 (November 2001): 785–796.

8 David E. Rowe, "Mathematical Models as Artefacts for Research: Felix Klein and the Case of Kummer Surfaces," *Mathematische Semesterberichte* 60 (2013): 5.

9 David Mumford, Caroline Series, and David Wright, *Indra's Pearls: The Vision of Felix Klein* (Cambridge: Cambridge University Press, 2002).

10 Stefan Halverscheid and Oliver Labs, "Felix Klein's Mathematical Heritage as Seen Through 3D Models," in *The Legacy of Felix Klein*, ed. Hans-Georg Weigand, William McCallum, Marta Menghini, Michael Neubrand, and Gert Schubring (Berlin: Springer, 2018), 132.

11 Halverscheid and Labs, "Felix Klein's Mathematical Heritage," 150.

12 Rowe, "Mathematical Models as Artefacts for Research," 6.

13 Herbert Mehrtens, "Mathematical Models," in *Models: The Third Dimension of Science*, ed. Soraya de Chadarevian and Nick Hopwood (Stanford: Stanford University Press, 2004), 294.

14 David E. Rowe, "Klein, Hilbert, and the Gottingen Mathematical Tradition," *Osiris* 5 (1989): 188.

15 In the twentieth century, it was not only the models themselves but also model-making techniques that were canonized, cataloged, and diffused. The model-making craft exemplified by the work of crystallographer René-Just Haüy and chemist Alexander Crum Brown was transmitted to twentieth-century architecture by several manuals, the most influential of which was H. M. Cundy and A. P. Rollett's 1951 *Mathematical Models*, which outlines in detail methods for assembling polytopic forms and hyperbolic shapes. Like Haüy's work, *Mathematical Models* appears repeatedly in the citation networks of core crystallographic designers and provided a critical nucleus of accessible scientific model-making. The book was widely embraced by designers and appeared in the bibliographies of many scientifically oriented architecture texts of the midcentury.

16 H. M. Cundy and A. P. Rollett, *Mathematical Models*, 2nd ed. (Oxford: Oxford University Press, 1961), 197.

17 Maria G. Bartolini Bussi, Daina Taimina, and Masami Isoda, "Concrete Models and Dynamic Instruments as Early Technology Tools in Classrooms at the Dawn of ICMI: From Felix Klein to Present Applications in Mathematics Classrooms in Different Parts of the World," *Mathematics Education* 42 (2010): 21.

18 Irene Polo-Blanco, "Theory and History of Geometric Models" (thesis, University of Groningen, 2007), 2.

19 Rowe, "Mathematical Models as Artefacts for Research," 17.

20 Peggy Aldrich Kidwell, Amy Ackerberg-Hastings, and David Lindsay Roberts, *Tools of American Mathematical Teaching, 1800–2000* (Baltimore: Johns Hopkins University Press, 2008), 221.

21 Peggy Kidwell, "Charter Members of the MAA and the Material Culture of American Mathematics," in *Research in History and Philosophy of Mathematics: The CSHPM 2015 Annual Meeting in Washington*, ed. Maria Zack and Elaine Landry (Basel: Birkhäuser, 2016), 214.

22 G. W. Cussons, "Mathematical Models," in *The Encyclopedia Britannica*, 14th ed., vol. 15 (London: Encyclopedia Britannica, 1929), 72.

23 Mark Burry, Jordi Coll Grifoll, and Josep Gómez Serrano, *Sagrada Família s. XXI: Gaudí ara/ahora/now* (Barcelona: Edicions UPC, 2008), 18.

24 Burry, Coll Grifoll, and Gómez Serrano, *Sagrada Família s. XXI*, 70.

25 C. W. Merrifield, "The Collection of Models of Ruled Surfaces at South Kensington," *Messenger of Mathematics* 3 (1874): 111–119.

26 College of New Jersey, *Catalog of the College of New Jersey at Princeton* (Princeton: College of New Jersey, 1889), 81.

27 *The Official Directory of the World's Columbian Exposition, May 1st to October 30th, 1893: A Reference Book of Exhibitors and Exhibits, and of the Officers and Members of the World's Columbian Commission* (Chicago: W. B. Conkey Company, 1893).

28 André Breton, *Exposition surréaliste d'objets* (Paris: Charles Ratton, 1936).

29 Quoted from "Au delà de la peinture," in *Max Ernst* (New York: Museum of Modern Art, 1961), 11.

30 Incidentally, Ernst introduced the singular collection of models at the Institut Henri Poincaré to Man Ray, who photographed them and, to some extent, popularized them among Paris-based artists. Jennifer Mundy, *Man Ray: Writings on Art* (Los Angeles: Getty Publications, 2016), 344.

31 Rowe, "Mathematical Models as Artefacts for Research," 5.

32 David Hilbert and Stephan Cohn-Vossen, *Geometry and the Imagination*, trans. Paul Nemenyi (New York: American Mathematical Society, 1952), iii.

33 Hilbert and Cohn-Vossen, *Geometry and the Imagination*, iv.

34 Robert Harbison, *Eccentric Spaces* (Cambridge, MA: MIT Press, 2000), 155.

35 Iannis Xenakis, "The Architectural Design of Le Corbusier and Xenakis," *Phillips Technical Review*, no. 1 (1958/59): 3.

36 Marc Treib, *Space Calculated in Seconds* (Princeton: Princeton University Press, 1996), 22.

37 Luca Sampò, "L'église Saint-Pierre de Firminy, de Le Corbusier: quarante ans d'histoire entre idée et réalisation," *Livraisons de l'histoire de l'architecture* 16 (2008): 2.

38 Herbert Read, "Constructivism: The Art of Naum Gabo and Antoine Pevsner" in *Gabo Pevsner* (New York: Museum of Modern Art. 1948), 11.

39 Ruth Olsen and Abraham Chanin, "Naum Gabo," in *Gabo Pevsner*, 15.

40 Olsen and Chanin, "Naum Gabo," 17.

41 Martin Hammer and Christina Lodder, *Constructing Modernity: The Art and Career of Naum Gabo* (New Haven: Yale University Press, 2000), 51.

42 Gabo documented many of his sculptures stereoscopically in *Gabo: Constructions, Sculpture, Paintings, Drawings, Engravings* (Cambridge, MA: Harvard University Press, 1957), 10.

43 Quoted in Hammer and Lodder, *Constructing Modernity*, 393.

44 Naum Gabo, "Art and Science," in *The New Landscape of Art and Science*, ed. György Kepes (Chicago: Paul Theobald, 1956), 61–63.

45 Leslie Martin, "Construction and Intuition," in Naum Gabo, *Gabo: Constructions, Sculpture, Paintings, Drawings, Engravings*, 9.

46 Leslie Martin, "The State of Transition," in *Circle: International Survey of Constructive Art*, ed. Leslie Martin, Ben Nicholson, and Naum Gabo (London: Faber & Faber, 1937), 217.

47 Antoine Pevsner, "Science Foils Poetry," *Leonardo* 10, no. 4 (Autumn 1977): 324–325. Based on an article published in French in *xxe Siecle* 12, no. 12 (May/June 1959):13.

48 Kunsthaus Zürich, *Antoine Pevsner, Georges Vantongerloo*, Max Bill (Zurich: Buchdruckerei Neue Zürcher Zeitung, 1949).

49 Max Bill, "The Mathematical Approach in Contemporary Art," in *Theories and Documents of Contemporary Art: A Sourcebook of Artists' Writings*, ed. Kristine Stiles and Peter Selz (Berkeley: University of California Press, 2012), 91.

50 Nicholas Bourbaki, "The Architecture of Mathematics," in *American Mathematical Monthly* 57, no. 4 (April 1950), 221–232

Chapter 4

1 Robin Evans, *The Projective Cast: Architecture and Its Three Geometries* (1995; repr., Cambridge, MA: MIT Press, 2000), 44.

2 Mario Carpo, *The Alphabet and the Algorithm* (Cambridge, MA: MIT Press, 2011), 54.

3 Matthew Edney, *Mapping an Empire: The Geographic Construction of British India, 1765–1843* (Chicago: University of Chicago Press, 1997), 293.

4 Buckminster Fuller, *Synergetics: Explorations in the Geometry of Thinking* (New York: Macmillan, 1975), 482.

5 Fokko Jan Dijksterhuis, "The Mutual Making of Sciences and Humanities: Willebrord Snellius, Jacob Goolius, and the Early Modern Entanglement of Mathematics and Philology," in *The Making of the Humanities*, vol. 2: *From Early Modern to Modern Disciplines* (Amsterdam: Amsterdam University Press, 2012), 75.

6 Edney, *Mapping an Empire*, 19.

7 M. Charvet, *L'art de la triangulation* (Grenoble: Peyronard, 1812).

8 Lorraine Daston, "Objectivity and the Escape from Perspective," *Social Studies of Science* 22 (1992): 607.

9 William Roy, "An Account of the Trigonometrical Operation, Whereby the Distance between the Meridians of the Royal Observatories of Greenwich and Paris Has Been Determined. By Major-General William Roy," *Philosophical Transactions of the Royal Society of London* 80 (1790), pl. III.

10 Roy, "An Account of the Trigonometrical Operation," 151.

11 D. Graham Burnett, *Masters of All They Surveyed: Exploration, Geography, and a British El Dorado* (Chicago: University of Chicago Press), 9.

12 Janis Langins, *Conserving the Enlightenment: French Military Engineering from Vauban to the Revolution* (Cambridge, MA: MIT Press, 2004), 47.

13 Sébastien Le Prestre de Vauban, *De l'attaque et de la défense des places* (The Hague: Pierre De Hondt, 1737).

14 Jean Picard, *Mesure de la terre* (Paris: L'Imprimere Royale, 1671).

15 Josef W. Konvitz, "The National Map Survey in Eighteenth-Century France," *Government Publications Review* 10 (1983): 395–403.

16 Ernst Breitenberger, "Gauss's Geodesy and the Axiom of Parallels," *Archive for History of Exact Sciences* 31, no. 3 (1984): 276.

17 Breitenberger, "Gauss's Geodesy and the Axiom of Parallels," 276.

18 Edney, *Mapping an Empire*, 242.

19 J. T. Walker, *Account of the Operations of the Great Trigonometrical Survey of India*, vol. 2 (Office of the Trigonometrical Branch, Survey of India, 1879).

20 Tarun Kumar Mondal, "Mapping India since 1767: Transformation from Colonial to Postcolonial Image," *Miscellanea Geographica* 23, no. 4 (2019), 211.

21 Edney, *Mapping an Empire*, 292.

22 G. Waldo Dunnington, *Carl Friedrich Gauss: Titan of Science* (New York: Hafner, 1955), 113.

23 Albrecht Meydenbauer, "Die Photometrographie," *Wochenblatt des Architektenvereins zu Berlin* 1, no. 14 (6 April, 1867): 125.

24 Jörg Albertz, "Albrecht Meydenbauer: Pioneer of Photogrammetric Documentation of the Cultural Heritage," in *Proceedings of the 18th International Symposium CIPA 2001*, ed. Jörg Albertz, François LeBlanc, and Christopher Gray (Berlin: CIPA 2001 Organising Committee, 2002), 20.

25 Miriam Paeslack, *Constructing Imperial Berlin: Photography and the Metropolis* (Minneapolis: University of Minnesota Press, 2019), 79.

26 Albertz, "Albrecht Meydenbauer," 22.

27 Fuller, *Synergetics*, 717.

28 Sigfried Giedion, *Space, Time and Architecture: The Growth of a New Tradition* (Cambridge, MA: Harvard University Press, 1941), 26.

29 Lagrange was in the circle of Parisian geometers which included his close contemporary Gaspard Monge, and later was one of the first professors at the École Polytechnique when Monge helped found it in 1794.

30 Erhard Scholz, "The Concept of Manifold, 1850–1950," in *History of Topology*, ed. I. M. James (Amsterdam: Elsevier, 1999), 25.

31 August Ferdinand Möbius, *Der barycentrische Calcül: ein neues Hülfsmittel zur analytischen Behandlung der Geometrie* (Leipzig: Barth, 1827).

32 Bernhard Riemann, "On the Hypotheses which Lie at the Foundation of Geometry," in *From Kant to Hilbert: A Source Book in the Foundations of Mathematics*, vol. 2, ed. William Ewald (Oxford: Oxford University Press, 2007), 652.

33 One particular responsibility of this group was to develop a proposal for a decimalization of time. Though there were several competing proposals, Poincaré proposed the division of the day by 400, providing a simple method of angle conversion. Jeremy Gray, *Henri Poincaré: A Scientific Biography* (Princeton: Princeton University Press, 2012), 188.

34 Gray, Henri Poincaré, 189.

35 Poincaré's thinking on the epistemology of science was broadly influential both within and outside scientific circles. He influenced his colleague Jacques Hadamard in his much later exposition of the subject in "An Essay on the Psychology of Invention in the Mathematical Field" (1949). But he also influenced those who hoped to further clarify hyperdimensionality in visual terms.

36 Esprit Jouffret, *Traité élémentaire de géométrie à quatre dimensions, et introduction à la géométrie à n dimensions* (Paris: Gauthier-Villars, 1903), xiii.

37 Jouffret, *Traité élémentaire de géométrie à quatre dimensions*, xv. Though he writes primarily in French, Jouffret employs a few English terms of art, including "blindfold-play."

38 William Irving Stringham. "Regular Figures in n-Dimensional Space," *American Journal of Mathematics* 3, no. 1 (March 1880): 1–14.

39 "Graduates," *The Harvard Register; an Illustrated Monthly* 2, no. 2 (August 1880): 166.

40 Quoted in Craig E. Adcock, *Marcel Duchamp's Notes from The Large Glass* (Ann Arbor: UMI Research Press, 1983), 64.

41 Alfred C. Lane, "Transcendental Geometry," *Popular Science Monthly* 21 (August 1, 1882): 507.

42 Quoted in Linda Dalrymple Henderson, *The Fourth Dimension and Non-Euclidean Geometry in Modern Art*, rev. ed. (Cambridge, MA: MIT Press, 2013), 448.

43 Antoine Pevsner, "Science Foils Poetry," *Leonardo* 10, no. 4 (Autumn 1977): 325.

44 Quoted in Pierre Cabanne, *Dialogues with Marcel Duchamp* (Cambridge: Da Capo Press, 2009), 23–24.

45 Henderson, *The Fourth Dimension and Non-Euclidean Geometry in Modern Art*, 13.

46 Claude Bragdon, *A Primer of Higher Space* (Rochester, NY: Manas Press, 1913), 1.

47 Hanoch Gutfreund and Jürgen Renn, *The Road to Relativity: The History and Meaning of Einstein's "The Foundation of General Relativity," Featuring the Original Manuscript of Einstein's Masterpiece* (Princeton: Princeton University Press, 2015), 105.

48 Iuliana Roxana Vicovanu, "L'Esprit Nouveau (1920–1925) and the Shaping of Modernism in the France of the 1920s" (PhD diss., Johns Hopkins University, 2009).

49 Henderson, *The Fourth Dimension and Non-Euclidean Geometry in Modern Art*, 463.

50 Paul Le Becq, "A propos des théories d'Einstein, élaboration d'une esthétique experimentale," *L'Esprit Nouveau*, no. 7 (1921): 719.

51 Le Becq, "A propos des théories d'Einstein," 727.

52 Henri Poincaré, "Pourquoi l'espace à trois dimensions," *De Stijl* 5, no. 5 (1923): 66–70.

53 Theo van Doesburg, "Towards a Plastic Architecture," in *De Stijl*, ed. Hans L. C. Jaffé (New York: H. N. Abrams), 185–188. Translated from Theo van Doesburg, "Tot Een Beeldende Architectuur," *De Stijl* 6, no. 6/7 (1924): 78–83.

54 Theo van Doesburg, "L'evolution de l'architecture moderne," *L'Architecture Vivant* 3, no. 9 (1925): 14–20.

55 Theo van Doesburg, "Elementarism (The Elements of the New Painting)," *Abstraction-création: art non-figuratif* no. 1 (1932): 39.

56 Nikolai Krasil'nikov, "Problems of Contemporary Architecture," *Sovremennaya Arkhitektura*, no. 6 (1928): 170–176, translated in Catherine Cooke, "Nikolai Krasil'nikov's Quantitative Approach to Architectural Design: An Early Example," *Environment and Planning B* 2 (1975): 3–20.

57 Krasil'nikov, "Problems of Contemporary Architecture."

58 L. Komarova and N. Krasil'nikov, "A Method of Investigating the Generation of Building Form," *Sovremennaya Arkhitektura*, no. 5 (1929), quoted in Lionel March, *The Architecture of Form* (Cambridge: Cambridge University Press, 1976), vii.

59 Katie Taylor, "Vernacular Geometry: Between the Senses and Reason," *BSHM Bulletin: Journal of the British Society for the History of Mathematics* 26, no. 3 (2011): 147–159.

Chapter 5

1 Jonathan Crary, *Techniques of the Observer* (Cambridge, MA: MIT Press, 1990).

2 Charles Wheatstone, "Contributions to the Physiology of Vision.— Part the First. On Some Remarkable, and Hitherto Unobserved, Phenomena of Binocular Vision," *Philosophical Transactions of the Royal Society of London* 128 (1838): 377.

3 Wheatstone, "Contributions to the Physiology of Vision," 380.

4 Wheatstone's priority in the development of stereoscopic drawing was not without debate; his bitter rival, David Brewster, erroneously argued that the Italian painter Jacopo Chimenti had in fact drawn stereoviews over two centuries earlier. Of course, the principle that each eye sees a slightly different image had long been discussed in optical circles, but Wheatstone's drawings were the first constructed images to achieve binocular fusion in full relief.

5 Oliver Wendell Holmes, "The Stereoscope and the Stereograph," *Atlantic*, June 1859, 738–748.

6 "The Stereoscope, Pseudoscope, and Solid Daguerreotypes," *Illustrated London News* (January 24, 1852): 78.

7 Thomas L. Hankins and Robert J. Silverman, *Instruments of the Imagination* (Princeton: Princeton University Press, 1995), 153.

8 "The Stereoscope, Pseudoscope, and Solid Daguerreotypes," 77.

9 Holmes, "The Stereoscope and the Stereograph."

10 Peter Galison, "Judgement against Objectivity," in *Picturing Science, Producing Art*, ed. Caroline Jones and Peter Galison (New York: Routledge, 1998), 332.

11 Jill Steward, "'How and Where To Go': The Role of Travel Journalism in Britain and the Evolution of Foreign Tourism, 1840–1914," in *Histories of Tourism: Representation, Identity and Conflict*, ed. John K. Walton (Bristol, UK: Channel View Publications, 2005).

12 Sue Beeton, *Travel, Tourism, and the Moving Image* (Bristol, UK: Channel View Publications, 2015), 46.

13 Robert DeLeskie, "The Underwood Stereograph Travel System: A Historical and Cultural Analysis" (master's thesis, Concordia University, 2001).

14 Hankins and Silverman, *Instruments of the Imagination*, 175.

15 H. Clay Price, "Stereoscopic," *Photographic Times* 33, no. 10 (October 1, 1901): 464.

16 "Stereoscopic Photographs: The Application of Stereoscopy to Clinical Records," *British Medical Journal* 2, no. 1979 (December 3, 1898): 1697–1698.

17 D. J. Cunningham and David Waterston, *The Edinburgh Stereoscopic Atlas of Anatomy* (Edinburgh: Jack, 1906).

18 Lorraine Daston and Elizabeth Lunbeck, "Observation Observed," in *Histories of Scientific Observation* (Chicago: University of Chicago Press, 2011), 7.

19 Étienne-Jules Marey, *Movement*, trans. Eric Pritchard (New York: D. Appleton, 1895), 24. Translated from Étienne-Jules Marey, *Le mouvement* (Paris: Librarie de l'Académie de Médecine, 1894).

20 Marey, *Movement*, 24.

21 Richard Difford, "In Defence of Pictorial Space: Stereoscopic Photography and Architecture in the Nineteenth Century," in *Camera Constructs: Photography, Architecture and the Modern City*, ed. Andrew Higgott and Timothy Wray (London: Routledge, 2016), 302.

22 Louis Albert Necker, "Observations on Some Remarkable Optical Phaenomena Seen in Switzerland; and on an Optical Phaenomenon Which Occurs on Viewing a Figure of a Crystal or Geometrical Solid," *London and Edinburgh Philosophical Magazine and Journal of Science* 1, no. 5 (1832): 336.

23 *Proceedings of the London Mathematical Society* 2 (1868): 58.

24 Felix Klein, Development of Mathematics in the 19th Century, vol. 9 (Brookline, MA: Math Sci Press, 1979), 229.

25 Klaus Hentschel, *Visual Cultures in Science and Technology: A Comparative History* (Oxford: Oxford University Press, 2014), 24.

26 Richard von Mises and Max von Laue, *Stereoskopbilder von Kristallgittern* (Berlin: J. Springer, 1926).

27 William H. Bragg and William L. Bragg, *Stereoscopic Photographs of Crystal Models* (London: Adam Hilger, 1928).

28 R. W. James, "Review: Stereoscopic Drawings of Crystal Structures," *Journal of Physical Chemistry* 40, no. 7 (1936): 934–935.

29 J. Bartels, "Geophysical Stereograms," *Terrestrial Magnetism and Atmospheric Electricity* 35 (1931): 187.

30 David Brewster, *The Stereoscope: Its History, Theory and Construction, with Its Application to the Fine and Useful Arts and to Education* (London: John Murray, 1856), 199.

31 Gilbert Pass, "The Stereoscope in Education," *The School World* 5 (1900): 278.

32 Edward Mann Langley, *Solid Geometry through the Stereoscope: Demonstrations of Some of the More Important Propositions* (Underwood & Underwood, 1907).

33 Penelope Haralambidou, *Marcel Duchamp and the Architecture of Desire* (London: Taylor and Francis, 2012), 153.

34 John T. Rule, "Stereoscopic Drawings," *Journal of the Optical Society of America* 28, no. 8 (1938): 313–322.

35 Rule, "Stereoscopic Drawings."

36 Rule, "Stereoscopic Drawings."

37 John T. Rule, "Apparatus for Producing Stereographic Drawings," US Patent 2,171,894 A, filed November 17, 1937, and issued September 5, 1939.

38 Sigfried Giedion, *Walter Gropius: Work and Teamwork* (New York: Reinhold, 1954).

39 Charles E. Benham, "Solid Curves with the Twin-Elliptic Pendulum," *Science Progress in the Twentieth Century* 22, no. 88 (April 1928): 607–612.

40 Charles E. Benham, "Descriptive and Practical Details as to Harmonographs," in *Harmonic Vibrations and Vibrational Figures*, ed. H. C. Newton (London: Newton and Co., 1909).

41 Benham, "Solid Curves with the Twin-Elliptic Pendulum," 612.

42 Michael Century, "Exact Imagination and Distributed Creativity: A Lesson from the History of Animation," *Proceedings of the 6th ACM SIGCHI Conference on Creativity and Cognition*, ed. Ben Shneiderman (New York: ACM, 2007), 83–90.

43 Century, "Exact Imagination and Distributed Creativity," 84.

44 Norman McLaren, "Stereographic Animation: The Synthesis of Stereoscopic Depth with Flat Drawing and Artwork," *Journal of the Society of Motion Picture and Television Engineers* 57, no. 6 (July–December 1951): 513.

45 Ray Zone, *3-D Revolution: The History of Modern Stereoscopic Cinema* (Lexington: University of Kentucky Press, 2012), 289.

46 Zone, *3-D Revolution*, 289.

47 McLaren, "Stereographic Animation," 513.

48 Carroll K. Johnson, OR-TEP: A FORTRAN Thermal-Ellipsoid Plot Program for Crystal Structure Illustrations (Oak Ridge, TN: Oak Ridge National Laboratory, 1965).

49 Johnson, OR-TEP, 1.

50 Alan H. Schoen, *Infinite Periodic Minimal Surfaces without Self-Intersections* (Washington, DC: NASA, 1970).

51 Schoen, *Infinite Periodic Minimal Surfaces*, 6.

52 Ivan Sutherland, "A Head-Mounted Three Dimensional Display," in *International Workshop on Managing Requirements Knowledge* (San Francisco, 1968).

53 Portions of the research for the project were completed at Lincoln Laboratories, MIT, Harvard, and the University of Utah.

54 Sutherland, "A Head-Mounted Three Dimensional Display," 757.

55 Sutherland, "A Head-Mounted Three Dimensional Display," 759.

56 Sutherland, "A Head-Mounted Three Dimensional Display," 759.

57 Sutherland, "A Head-Mounted Three Dimensional Display," 760.

58 Sutherland, "A Head-Mounted Three Dimensional Display."

59 Sutherland, "A Head-Mounted Three Dimensional Display," 757 (my emphasis).

60 Robert Harbison, *Eccentric Spaces* (Cambridge, MA: MIT Press, 2000), 74.

61 Robert J. Silverman, "The Stereoscope and Photographic Depiction in the 19th Century," *Technology and Culture* 34, no. 4 (October 1993): 729–756.

Chapter 6

1 Charles Thomson Rees Wilson, "On a Method of Making Visible the Paths of Ionising Particles through a Gas," *Proceedings of the Royal Society A* 85, no. 578 (1911): 285.

2 Peter Galison and Alexi Assmus, "Artificial Clouds, Real Particles," in *The Uses of Experiment: Studies in the Natural Sciences* (Cambridge: Cambridge University Press, 1989), 226.

3 Colin Williamson, "Quicker than the Eye: Science, Cinema, and the Question of Vision," in *Hidden in Plain Sight: An Archaeology of Magic and the Cinema* (New Brunswick: Rutgers University Press, 2015), 50.

4 László Moholy-Nagy, *The New Vision: Fundamentals of Design, Painting, Sculpture, Architecture* (New York: W. W. Norton, 1938), 19.

5 Moholy-Nagy, *New Vision*, 19.

6 Peter Burke, *A Social History of Knowledge*, vol. 2 (Cambridge, UK: Polity, 2012), 162, 172.

7 *A History of the Cavendish Laboratory, 1871–1910* (London: Longmans, Green and Co., 1910), 2.

8 Josep Simon, *Communicating Physics: The Production, Circulation, and Appropriation of Ganot's Textbooks in France and England, 1851–1887* (Pittsburgh: University of Pittsburgh Press, 2016), 16.

9 Felix Auerbach, *Physik in graphischen Darstellungen* (Leipzig: Druck und Verlag von B. G. Teubner, 1912).

10 Felix Auerbach, *Die graphischen Darstellungen* (Leipzig: Druck und Verlag von B. G. Teubner, 1914).

11 Robert Brain, *The Pulse of Modernism: Physiological Aesthetics in Fin-de-Siècle Europe* (Seattle: University of Washington Press, 2015), 11.

12 Brain, *Pulse of Modernism*, 6.

13 V. J. McGill, "Logical Positivism and the Unity of Science," *Science and Society* 1, no. 4 (Summer 1937): 550.

14 Friedrich Stadler, "The 'Verein Ernst Mach'—What Was It Really?," in *Ernst Mach—A Deeper Look: Documents and Perspectives*, ed. John Blackmore, Boston Studies in the Philosophy of Science 143 (Springer: Dordrecht, 1992), 354.

15 Bertrand Russell and Alfred North Whitehead, *Principia Mathematica* (Cambridge: Cambridge University Press, 1910).

16 Albert E. Blumberg and Herbert Feigl, "Logical Positivism," *Journal of Philosophy* 28, no. 11 (May 21, 1931): 282.

17 George A. Reisch, "Planning Science: Otto Neurath and the 'International Encyclopedia of Unified Science,'" *British Journal for the History of Science* 27, no. 2 (June 1994): 153.

18 Otto Neurath, "Unified Science and Its Encyclopaedia," *Philosophy of Science* 4, no. 2 (April 1937): 266.

19 Neurath, "Unified Science and Its Encyclopaedia," 277.

20 Notably the IEUS was the first venue of publication of Thomas Kuhn's Structure of Scientific Revolutions, a seminal work at the intersection of the epistemology and sociology of science. Thomas S. Kuhn, *The Structure of Scientific Revolutions* (Chicago: University of Chicago Press, 1962); Charles Morris, "On the History of the International Encyclopedia of Unified Science," *Synthese* 12, no. 4 (December 1960): 520.

21 Peter Galison, "Aufbau/Bauhaus: Logical Positivism and Architectural Modernism," *Critical Inquiry* 16, no. 4 (Summer 1990): 709–752. The permeable boundary between positivistic philosophy and the Bauhaus was also reinforced by overlapping social circles. Galison noted that Carnap himself saw the Bauhaus as something of an analogue to logical positivism. Carnap and Neurath also lectured at the Bauhaus, and Charles Morris would later appear as a pivotal intellectual figure in László Moholy-Nagy's New Bauhaus. Eve Blau has noted that the relationship between the positivists and the Bauhaus was not uncritical, and Neurath, for one, ultimately became disillusioned with the Bauhaus. Yet the paths of positivists and designers seemed to converge toward a common project of epistemic unification.

22 Walter Gropius, "Reorientation," in *The New Landscape of Art and Science*, ed. György Kepes (Chicago: Paul Theobald, 1956), 94.

23 Walter Gropius, *Scope of Total Architecture* (New York: Collier, 1962), 30.

24 Walter Gropius, "In Search of a Common Denominator of Design Based on the Biology of the Human Being," Walter Gropius Papers, 1925–1969, MS Ger 208, file 93, Houghton Library, Harvard Special Collections, Harvard University.

25 Gropius, "In Search of a Common Denominator of Design."

26 Gropius, *Scope of Total Architecture*, 25.

27 László Moholy-Nagy, "Light Painting," in *Circle: International Survey of Constructive Art*, ed. Leslie Martin, Ben Nicholson, and Naum Gabo (New York: Praeger, 1971), 245.

28 Walter Gropius, "Optical Illusions to Be Considered in Design Training," Walter Gropius Papers, 1925–1969, MS Ger 208, file 36, Houghton Library, Harvard Special Collections, Harvard University.

29 Gropius, "Optical Illusions to Be Considered in Design Training."

30 Quoted in Howard Dearstyne, *Inside the Bauhaus* (London: Elsevier, 2014), 58.

31 Barry Bergdoll and Leah Dickerman, *Bauhaus 1919–1933: Workshops for Modernity* (New York: Museum of Modern Art, 2009), 30.

32 Moholy-Nagy, *New Vision*, 138.

33 Sibyl Moholy-Nagy, *Moholy-Nagy: Experiment in Totality* (Cambridge, MA: MIT Press, 1969), 129.

34 Beaumont Newhall, "The Photography of Moholy-Nagy," *Kenyon Review* 3, no. 3 (1941): 351.

35 Rosalind E. Krauss, *Passages in Modern Sculpture* (Cambridge, MA: MIT Press, 1981), 207.

36 Moholy-Nagy, *Moholy-Nagy: Experiment in Totality*, 170.

37 Joseph Malherek, "The Industrialist and the Artist: László Moholy-Nagy, Walter Paepcke, and the New Bauhaus in Chicago, 1918–46," *Journal of Austrian-American History* 2, no. 1 (2018): 51.

38 Charles Morris, "Science, Art, and Technology," *Kenyon Review* 1, no. 4 (1939): 409.

39 Quoted in Moholy-Nagy, *Moholy-Nagy: Experiment in Totality*, 153.

40 Williamson, "Quicker than the Eye," 51.

41 Jules Antoine Lissajous, *Mémoire sur l'étude optique des mouvements vibratoires* (Paris: Mallet-Bachelier, 1857).

42 Étienne-Jules Marey, *Movement*, trans. Eric Pritchard (New York: Appleton, 1895), 99.

43 Marey, *Movement*, 84–102.

44 Étienne-Jules Marey, *La méthode graphique dans les sciences expérimentales* (Paris: Librairie de l'Académie de Médecine, 1887), i (author's translation).

45 Frank Bunker Gilbreth and Lillian Moller Gilbreth, *Applied Motion Study: A Collection of Papers on the Efficient Method to Industrial Preparedness* (New York: Macmillan, 1919), 42–43.

46 Gilbreth and Gilbreth, *Applied Motion Study*, 89.

47 Frank and Lillian Gilbreth, "A Fourth Dimension for Measuring Skill for Obtaining the One Best Way to Do Work," *Society of Industrial Engineering Bulletin* 5, no. 11 (1923): 6.

48 Gilbreth and Gilbreth, *Applied Motion Study*, vii.

49 Gilbreth and Gilbreth, *Applied Motion Study*, 99.

50 Jane Callaghan and Catherine Palmer, *Measuring Space and Motion* (New York: John B. Pierce Foundation, 1944), 4.

51 Callaghan and Palmer, *Measuring Space and Motion*, 8.

52 Ronald R. Kline, *The Cybernetics Moment: Or Why We Call Our Age the Information Age* (Baltimore: Johns Hopkins University Press, 2015), 50.

53 "Astronauts Continue Tests in MASTIF," in *Orbit: NASA Lewis Research Center* 18, no. 5 (1960): 2.

54 Clovis Heimsath, *Behavioral Architecture: Toward an Accountable Design Process* (New York: McGraw-Hill, 1977).

55 *The Art of Projection and Complete Magic Lantern Manual* (London: E. A. Beckett, 1893), 156.

56 William Rigge, *Harmonic Curves* (Omaha: Creighton University, 1926).

57 Rigge, *Harmonic Curves*, 89.

58 Elaine O'Hanrahan, "The Contribution of Desmond Paul Henry (1921–2004) to Twentieth-Century Computer Art," *Leonardo* 51, no. 2 (2018): 157.

59 Kepes, *New Landscape of Art and Science*, 173.

60 The Walter Gropius Papers, 1925–1969, MS Ger 208, Houghton Library, Harvard Special Collections, Harvard University.

Chapter 7

1 Since Meinong had a close relationship to the development of Gestalt psychology, some of these entities were naturally geometric.

2 Dale Jacquette, *Alexius Meinong: The Shepherd of Non-Being* (Cham: Springer, 2015).

3 Jacquette, *Alexius Meinong*, xxi.

4 Brian Rotman, "Topology, Algebra, Diagrams, Theory," *Culture and Society* 29, no. 4 (July 2012): 248.

5 Ben van Berkel and Caroline Bos, "Urban Surfaces: O.C.E.A.N Net 96," *AA Files*, no. 33 (Summer 1997): 84.

6 Ben van Berkel and Caroline Bos, "Diagrams: Interactive Instruments in Operation," *Any* 23 (1998): 22.

7 Moritz Epple, "Topology, Matter, and Space, I: Topological Notions in 19th-Century Natural Philosophy," *Archive for History of Exact Sciences* 52 (1998): 322–325.

8 Peter Guthrie Tait, "On Knots," in *Scientific Papers* (Cambridge: Cambridge University Press, 1898), 304.

9 Alexander Crum Brown, "On a Case of Interlacing Surfaces," *Proceedings of the Royal Society of Edinburgh* 13 (1885–1886): 383.

10 Crum Brown was no stranger to the space between dimensions: he was already renowned for developing a two-dimensional nomenclature for three-dimensional chemical bonds, effectively unfolding intricate lattices into flat notation. Alexander Crum Brown, "On the Theory of Isomeric Compounds," *Transactions of the Royal Society of Edinburgh* 23 (1864): 707–719.

11 Tait, "On Knots," 298. Tait actually mentions Plateau by name, recommending his minimal-surface technique for knot constructions.

12 Joseph Plateau, *Statique expérimentale et théorique des liquides soumis aux seules forces moléculaires* (Leipzig: Clemm, 1873). This book was actually a summary of two decades of prior work, undertaken between 1843 and 1868.

13 As a physicist, Plateau had broad interests, being one of the first to create the illusion of motion with his phenakistiscope in 1832.

14 Kristel Wautier, Alexander Jonckheere, and Danny Segers, "The Life and Work of Joseph Plateau: Father of Film and Discoverer of Surface Tension," *Physics in Perspective* 14 (2012): 258–278.

15 Max Bill, *Surfaces* (Toronto: Marlborough Gallery, 1972), 3.

16 Max Bill, "The Mathematical Approach in Contemporary Art," in *Theories and Documents of Contemporary Art: A Sourcebook of Artists' Writings*, ed. Kristine Stiles and Peter Selz (Berkeley: University of California Press, 2012), 91.

17 Bill, *Surfaces*, 7.

18 David Hilbert and Stephan Cohn-Vossen, *Geometry and the Imagination*, trans. Paul Nemenyi (New York: American Mathematical Society, 1952).

19 Max Dehn and Poul Heegaard, "Analysis situs," in *Enzyklopädie der mathematischen Wissenschaften mit Einschluss ihrer Anwendungen*, vol. 3, ed. H. Burkhardt, M. Wirtinger, and R. Fricke (Leipzig: B. G. Tuebner, 1907), 153–220.

20 Philip Ording, "A Definite Intuition," *Bulletins of the Serving Library* 5 (2013).

21 *Black Mountain College Bulletin / Bulletin-Newsletter* 3, no. 7 (July 1945): 8.

22 Max Dehn Papers, 1899–1979, Box 4RM132, University of Texas at Austin.

23 Ording, "A Definite Intuition."

24 Max Wertheimer, "Laws of Organization in Perceptual Forms," in *A Source Book of Gestalt Psychology*, ed. W. D. Ellis (London: Routledge & Kegan, 1938), 71–88.

25 Max Wertheimer, "Laws of Organization in Perceptual Forms" in *A Source Book of Gestalt Psychology*, vol. 2, ed. Willis D. Ellis (New York: Psychology Press, 1999).

26 Roy R. Behrens, "Art, Design and Gestalt Theory," *Leonardo* 31, no. 4 (August 1998): 299–303.

27 Klee also explored the implied superposition and transparency evoked by Wertheimer in a distinct set of his drawings and paintings. His *Uncomposed Objects in Space* (1929), for example, was an abstract wireframe vision of existence which trades in gestalt idioms of polyvalent spatial wireframes.

28 Anthony W. Auerbach, "Structural Constellations: Excursus on the Drawings of Josef Albers c. 1950–1960" (PhD diss., University College London, 2003).

29 The famous illusion developed by the crystallographer Louis Albert Necker in 1832.

30 Auerbach, "Structural Constellations."

31 Josef Albers, *The Interaction of Color* (New Haven: Yale University Press, 2013), 2.

32 Bruno Petermann, *Das Gestaltproblem in der Psychologie im Lichte analytischer Besinnung. Ein Versuch zu grundsätzlicher Orientierung* (Leipzig: J. A. Barth, 1931).

33 Matila Ghyka, *The Geometry of Art and Life* (New York: Sheed and Ward, 1946).

34 Hermann Weyl, *Symmetry* (Princeton: Princeton University Press, 1952).

35 Walter Lietzmann, *Mathematik und bildende Kunst* (Breslau: Ferdinand Hirt, 1931). One of Lietzmann's best known books, Anschauliche Topologie, was a thoroughly visual introduction to topology, along the lines of Felix Klein's pedagogy.

36 An intriguing later addition was Kurd Alsleben's *Drei Probleme aus dem Bereich der Informationsästhetik* (Munich: Dieter Hacker and Klaus Staudt, 1966), which explicitly treated so-called computer art.

37 Herbert Seifert, "Über das Geschlecht von Knoten," *Mathematische Annalen* 110 (1935): 572.

38 Engman later went on to head Yale's sculpture program. His working methodology embraced precise geometric transformations as well as an intuitive resolution of the resulting complexity.

39 The other artists represented at the show were John Cunningham, Deborah de Moulpied, William Reimann, Stephanie Scuris, and Robert Zeidman.

40 Josef Albers, "Structural Sculpture," in *Structural Sculpture* (New York: Galerie La Chalette), 1961.

41 Albers hired Hauer to be a member of the Yale sculpture faculty, where Hauer's mathematical interests ranged widely in his courses. Beyond the consideration of his empirical topology, he introduced students to tensegrity structures, projections of the hypercube, and precise methods of partitioning space. Erwin Hauer, interview by author, New Haven, 2012.

42 Henry Moore: *Writings and Conversations*, ed. Alan Bowness (Berkeley: University of California Press, 2002), 256.

43 For example, "New Forms in Sculptural Walls and Facings," *Progressive Architecture* 42, no. 6 (June 1961): 18; "Light Diffusing Walls," *Progressive Architecture* 46, no. 7 (July 1965): 74; "Sculptured Walls," *Progressive Architecture* 47, no. 3 (March 1966): 87–89.

44 Hauer, interview by author.

45 Hauer, interview by author.

46 Daniela Fabricius, "Material Models, Photography, and the Threshold of Calculation," *Arq* 21, no. 1 (2017): 21–32.

47 Cornelie Leopold, "Precise Experiments: Relations between Mathematics, Philosophy and Design at Ulm School of Design," *Nexus Network Journal* 15 (2013): 371.

48 Rene Spitz, *The Ulm School of Design: A View behind the Foreground* (Stuttgart: Edition Axel Menges, 2002), 219.

49 Heiner Jacob, "HfG Ulm: A Personal View of an Experiment in Democracy and Design Education," *Journal of Design History* 1, no. 3/4 (1988): 233.

50 Tomás Maldonado and Gui Bonsiepe, "Science and Design," *Ulm Journal* 10–11 (May 1965): 14.

51 Paul Betts, *The Authority of Everyday Objects: A Cultural History of West German Industrial Design* (Berkeley: University of California Press, 2007), 143.

52 Gui Bonsiepe and John Cullars, "The Invisible Facets of the Hfg Ulm," *Design Issues* 11, no. 2 (1995): 16.

53 Max Bill, "Beauty from Function and as Function," in *Idea* 53, ed. Gerd Hatje (New York: Wittenborn Shultz, 1952), x.

54 Lech Tomaszewski, "Regular Forms of Closed Non-orientable Surfaces," *Situationist Times* 5 (1964): 13–14.

55 Ron Resch, "The Topological Design of Sculptural and Architectural Systems," in *Proceedings of the June 4–8, 1973, National Computer Conference and Exposition, New York, New York* (New York: Association of Computing Machinery, 1973), 643–650.

56 Resch, "Topological Design of Sculptural and Architectural Systems," 650.

57 Resch, "Topological Design of Sculptural and Architectural Systems," 645.

58 Jean des Cars and Pierre Pinon, *Paris: Haussmann* (Paris: Editions Pavillion de l'Arsenal, 1991), 300.

59 John Desmond Bernal, "Art and the Scientist," in *Circle: International Survey of Constructive Art*, ed. Leslie Martin, Ben Nicholson, and Naum Gabo (London: Faber and Faber, 1937), 119–123.

60 John Desmond Bernal, "Architecture and Science," in *The Freedom of Necessity* (London: Routledge, 1949), 194. Originally published in Journal of the Royal Institute of British Architects, no. 16 (June 1937).

61 Robert Le Ricolais, "Topology and Architecture," *Student Publications of the School of Design of North Carolina State College* 5, no. 2 (1955): 10–16.

62 "Studies in the Functions and Design of Hospitals (Review)," *Proceedings of the Royal Society of Medicine* 49, no. 5 (May 1956): 300.

63 The Nuffield Trust, *Studies in the Functions and Design of Hospitals* (London: Oxford University Press, 1955), xix, 6.

64 Gerald L. Delon and Harold Eugene Smalley, *Quantitative Methods for Evaluating Hospital Designs* (Rockville, MD: National Center for Health Services, Research, and Development, 1970), 10.

65 Marc J. de Vries, Nigel Cross, and D. P. Grant, eds., *Design Methodology and Relationships with Science* (Dordrecht: Springer Science & Business Media, 2013).

66 Lionel March and Philip Steadman, *The Geometry of the Environment* (Cambridge, MA: MIT Press, 1971).

67 S.L.R., "Designer's Utopia?," *Progressive Architecture* 52, no. 7 (July 1971): 87.

68 S.L.R., "Designer's Utopia?," 87.

69 Mark Wigley, "Network Fever," *Grey Room*, no. 4 (Summer 2001): 91.

70 Stephen Jay Gould and R. C. Lewontin, "The Spandrels of San Marco and the Panglossian Paradigm: A Critique of the Adaptationist Programme," *Proceedings of the Royal Society of London. Series B, Biological Sciences* 205, no. 1161 (1979): 581–598.

71 Ranko Bon, "An Introduction to Morphometric Analysis of Spatial Phenomena on Micro-Environmental Scale" (master's thesis, Harvard University, 1972).

72 Ranko Bon, "Allometry in the Topologic Structure of Architectural Spatial Systems," *Ekistics* 36, no. 215 (October 1973): 270–276.

73 Philip Steadman, "Allometry and Built Form: Revisiting Ranko Bon's Work with the Harvard Philomorphs," *Construction Management and Economics* 24, no. 7 (2006): 755–765.

74 J. N. Findlay, *Meinong's Theory of Objects and Values* (Oxford: Clarendon Press, 1963), 47.

75 Findlay, *Meinong's Theory of Objects and Values*, 49.

76 March and Steadman, *Geometry of the Environment*, 194.

77 L. S. Penrose and R. Penrose, "Impossible Objects: A Special Type of Visual Illusion," *British Journal of Psychology* 49, no. 1 (February 1958): 31.

78 D. N. Perkins, "Gestalt Theory Is Alive and Well and Living in Information-Processing Land: A Response to Arnheim," *New Ideas in Psychology* 4, no. 3 (1986): 295–299.

79 D. A. Huffman, "Impossible Objects as Nonsense Sentences," in *Machine Intelligence* 6, ed. Bernard Meltzer and Donald Michie (Edinburgh: Edinburgh University Press, 1971).

Chapter 8

1 Ironically, the Rubik's cube was invented by an architect, Ernő Rubik of Hungary.

2 William Davies, "Introduction to Economic Science Fictions," in *Economic Science Fictions*, ed. William Davies (London: Goldsmith's Press, 2018), 13.

3 Pier Vittorio Aureli, "Appropriation, Subdivision, Abstraction: A Political History of the Urban Grid," *Log* 44 (Fall 2018): 139.

4 Andrew L. Russell, "Modularity: An Interdisciplinary History of an Ordering Concept," *Information and Culture: A Journal of History* 47, no. 3 (2012): 262.

5 Federated American Engineering Societies, *Waste in Industry* (New York: McGraw-Hill, 1921), 54.

6 Federated American Engineering Societies, *Waste in Industry*, 59.

7 Federated American Engineering Societies, *Waste in Industry*, 90.

8 Albert Farwell Bemis, *The Evolving House*, vol. 3: *Rational Design* (Cambridge, MA: Massachusetts Institute of Technology, 1934), 5.

9 Bemis, *Evolving House*, vol. 3, 585.

10 Bemis, *Evolving House*, vol. 3, 403.

11 Bemis, *Evolving House*, vol. 3, 41.

12 Bemis, *Evolving House*, vol. 3, 66.

13 Bemis, *Evolving House*, vol. 3, 194.

14 Bemis, *Evolving House*, vol. 3, 123.

15 Bemis, *Evolving House*, vol. 3, 43.

16 Bemis, *Evolving House*, vol. 3, 17.

17 Bemis, *Evolving House*, vol. 3, 17.

18 Myron Whitlock Adams and Prentice Bradley, *A62 Guide for Modular Coordination: A Guide to Assist Architects and Engineers in Applying Modular Coordination to Building Plans and Details* (Modular Service Association, 1946).

19 United States Housing and Home Finance Agency, *Basic Principles of Modular Coordination* (Washington, DC: United States Housing and Home Finance Agency, 1953), 4.

20 "Expansible Prefab House for Postwar," *Architectural Record* 96, no. 6 (December 1944): 69.

21 United States Housing and Home Finance Agency, International Housing Agency, "Review of Modular Measure," *Ekistics* 6, no. 38 (December 1958): 331–333.

22 Martin had a longstanding commitment to objective processes in design. He had coedited, with Naum Gabo and Ben Nicholson, the 1937 *Circle: International Survey of Constructive Art*, which anthologized modern tendencies in art and design for a British audience.

23 John Stanley Durrant Bacon, *The Chemistry of Life* (London: Watts & Company, 1944). The incident is recounted in Adam Sharr and Stephen Thornton, *Demolishing Whitehall: Leslie Martin, Harold Wilson and the Architecture of White Heat* (Surrey: Ashgate, 2013).

24 Sharr and Thornton, *Demolishing Whitehall*, 172.

25 Leslie A. Martin and Lionel March, eds., *Urban Space and Structures* (Cambridge: Cambridge University Press, 1972), 54.

26 Martin and March, *Urban Space and Structures*, 33.

27 Leslie Martin, "Architects' Approach to Architecture," *Journal of the Royal Institute of British Architects* 74 (1967): 191–200.

28 Leslie Martin, "The Grid as Generator," in Martin and March, *Urban Space and Structures*, 6–27.

29 Martin, "The Grid as Generator."

30 Marc Levinson, *The Box: How the Shipping Container Made the World Smaller and the World Economy Bigger* (Princeton: Princeton University Press, 2008).

31 Levinson, *The Box*.

32 Matthew W. Heins, "The Shipping Container and the Globalization of American Infrastructure" (PhD diss., University of Michigan, 2013), 20.

33 Centrum voor Cubische Constructies, *Cubic Constructions Compendium* (Deventer, Netherlands: Octopus Foundation, 1970), 244.

34 Centrum voor Cubische Constructies, *Cubic Constructions Compendium*, 3.

35 Centrum voor Cubische Constructies, *Cubic Constructions Compendium*, 8.

36 Centrum voor Cubische Constructies, *Cubic Constructions Compendium*, 7.

37 Sean Keller, "Fenland Tech: Architectural Science in Postwar Cambridge," *Grey Room*, no. 23 (Spring 2006): 40–65.

38 Lionel March and Philip Steadman, *The Geometry of Environment: An Introduction to Spatial Organization in Design* (London: RIBA Publications, 1971), 200.

39 March and Steadman, *Geometry of Environment*, 217.

40 March and Steadman, *Geometry of Environment*, 217.

41 Reinhold Martin, "Atrocities: Or, Curtain Wall as Mass Medium," *Perspecta* 32 (2001): 66–75.

42 Martin, "Atrocities," 67.

43 Martin, "Atrocities," 67.

44 For example, in James G. Colsher, "Iterative Three-Dimensional Image Reconstruction from Tomographic Projections," *Computer Graphics and Image Processing* 6, no. 6 (December 1977): 513–537.

45 Yona Friedman, "The Flatwriter: Choice by Computer," *Progressive Architecture* 52, no. 3 (1971): 4.

46 Friedman, "The Flatwriter," 4.

47 Friedman, "The Flatwriter," 99.

48 Yona Friedman, *Towards a Scientific Architecture* (Cambridge, MA: MIT Press, 1975), 53.

49 Friedman, "The Flatwriter," 99.

50 Yona Friedman, "l'urbanisme spatial," *Architecture, Formes, Fonctions* 12 (Lausanne: Anthony Krafft, 1965), 66. This text includes the English translation.

51 Friedman, "L'urbanisme spatial," 68; "une infrastructure a l'échelle du monde" (my translation).

52 Friedman, "The Flatwriter," 99.

53 Richard Thaler, Cass Sunstein, and John Balz, "Choice Architecture," *SSRN Electronic Journal* (2010).

54 Friedman, *Towards a Scientific Architecture*, 44.

55 Friedman, *Towards a Scientific Architecture*, 27.

56 Friedman, *Towards a Scientific Architecture*, 81.

57 Friedman, *Towards a Scientific Architecture*, 102.

58 Friedman, *Towards a Scientific Architecture*, 59.

59 Friedman, *Towards a Scientific Architecture*, 8.

60 Friedman, "The Flatwiter," 98.

61 Roberto Bottazzi, *Digital Architecture beyond Computers* (London: Bloomsbury, 2018), 68.

62 R. Buckminster Fuller, *Earth, Inc.* (New York: Anchor, 1973), 179.

63 R. Buckminster Fuller, "The World Game," *Ekistics* 28, no. 167 (October 1969): 287.

64 Fuller, "The World Game," 287.

65 Fuller, "The World Game," 291.

66 R. Buckminster Fuller, "10 Proposed Solutions for Improving World," *New York Times*, June 29, 1972, 41.

Chapter 9

1 Thomas Creighton, "The New Sensualism," *Progressive Architecture* 40, no. 9 (September 1, 1959): 141.

2 Beatriz Colomina, *Privacy and Publicity: Modern Architecture as Mass Media* (Cambridge, MA: MIT Press, 1994), 14.

3 E. Levin, "The Shape of Roofs to Come," *Manchester Guardian*, June 10, 1959, 14.

4 Christopher Wren, "Engine Plus Figures I—VI.," *Philosophical Transactions of the Royal Society* 4, no. 53 (1669).

5 William Crosby Marshall, *Descriptive Geometry for the Use of Students in Engineering*, vol. 1 (New Haven: William C. Marshall, 1909), 94.

6 Шухов, http://vystavki.rgantd.ru/shuhov/, accessed June 7, 2020.

7 Elizabeth Cooper English, "Arkhitektura i Mnimosti: The Origins of Soviet Avant-Garde Rationalist Architecture in the Russian Mystical Philosophical and Mathematical Intellectual Tradition" (PhD diss., University of Pennsylvania, 2000), 20.

8 English, "Arkhitektura i Mnimosti," 25.

9 Terri Meyer Boake, *Diagrid Structures: Systems, Connections, Details* (Berlin: De Gruyter, 2014), 20.

10 English, "Arkhitektura i Mnimosti," 25.

11 Stephen P. Timoshenko, "The Development of Engineering Education in Russia," *Russian Review* 15, no. 3 (July 1956): 174.

12 Timoshenko, "Development of Engineering Education in Russia," 173.

13 Rene Motro and Bernard Maurin, "Bernard Laffaille, Nicolas Esquillan, Two French Pioneers," in *IASS-IABSE Symposium: Taller, Longer, Lighter* (Zurich: IABSE / IASS, 2011), 8. For example, Bernard Laffaille, "Nouvelle surface en béton armé," FR Patent FR763842A, filed January 31, 1931, and issued May 7, 1934; Bernard Laffaille, "Perfectionnements dans la construction des voûtes," FR Patent FR802249A, filed May 5, 1935, and issued August 31, 1936; Bernard Laffaille, "Nouvelle surface de construction," FR799443A, filed March 14, 1935, and issued June 12, 1936.

14 Michele Melaragno, *An Introduction to Shell Structures: The Art and Science of Vaulting* (New York: Springer, 1991); C. Greco, "Giorgio Baroni, coperture sottili in cemento armato in forma di paraboloide iperbolico," *Area* 57, no. 12 (2001): 24–31.

15 Iasef M. Rian and Mario Sassone, "Tree-Inspired Dendriforms and Fractal-Like Branching Structures in Architecture: A Brief Historical Overview," *Frontiers of Architectural Research* 3, no. 3 (September 2014): 298–323.

16 For example, Giorgio Baroni, "Improvements in and Relating to Reinforced Concrete Roofs and the Like and to a Process of Manufacture of Same," GB Patent GB505787A, filed November 23, 1937, issued May 17, 1939.

17 Konrad Hruban, "Some Recent Shell Structures," *Architectural Science Review* 6, no. 2 (1963): 44–49.

18 Melaragno, *Introduction to Shell Structures*, 146.

19 Eduardo Torroja, "Un nuevo pavimento," ES Patent ES111876A1, filed March 13, 1929, and issued May 16, 1929; Eduardo Torroja, "Un procedimiento de construir estructuras resistentes superficiales, formadas por una triangulación de piezas metalicas," ES Patent ES184967A1, filed August 19, 1948, and issued March 16, 1949. Additional patents include ES108906A1, ES111877A1, ES118282A1, ES129993A1, ES145952A1, and ES180204A1.

20 Felix Candela, "Understanding the Hyperbolic Paraboloid," *Architectural Record* 123, no. 7 (July 1958): 191–195.

21 Tyler S. Sprague, "Beauty, Versatility, Practicality: The Rise of Hyperbolic Paraboloids in Post-war America (1950–1962)," *Construction History* 28, no. 1 (2013): 165–184.

22 Sprague, "Beauty, Versatility, Practicality," 170.

23 "Tomorrow's Life Today," *Life*, November 11, 1957, 132.

24 "New Shapes of Shelter," *Life*, November 11, 1957, 136.

25 "New Shapes of Shelter," 135.

26 Eduardo Catalano, *The Constant: Dialogues on Architecture in Black and White* (Cambridge, MA: Cambridge Architectural Press, 2000).

27 Occasionally he even includes biological references, speculations on the spatial taxonomies that Linnaeus or Darwin would observe in architecture. Nevertheless, these biological analogies are couched in distinctly objective terms, suggesting that the rigor of mathematics is never far from Catalano's view of the sciences.

28 Nowicki submitted no patents, while Catalano submitted one for a chair design.

29 Robin Evans, *The Projective Cast* (1995; repr., Cambridge, MA: MIT Press, 2000), 348. Of course, Le Corbusier had long been fascinated by mathematics and the precision it allegedly afforded designers, invoking an almost mythic reverence for geometry: "Relying on calculations, engineers use geometric forms, satisfying our eyes through geometry and our minds through mathematics; their works are on the way to great art" (Le Corbusier, *Toward an Architecture*, trans. John Goodman [Los Angeles: Getty Research Institute, 2007], 85). *Toward an Architecture* has several striking examples of what Le Corbusier called the "intense joys of geometry," surfaces which were argued as perfectly optimal to specific functions, such as propellers, airplane wings, and boat hulls. The shape of perfectly tuned performance, it seems, seduced the designer. He waxed poetic: "Architecture is plastic invention, is intellectual speculation, is higher mathematics" (Le Corbusier, *Toward an Architecture*, 189).

30 "General Motors Technical Center: Design Elements—Research Building Stairs," *Progressive Architecture* 36, no. 2 (February 1955): 102.

31 E. Levin, *Manchester Guardian*, June 10, 1959, 14.

32 Hugh J. Wylie, "Professor's New Home Is Mostly Two Roofs," *Washington Post* and *Times Herald*, January 19, 1957, B6.

33 "Hyperbolic Paraboloid Gets First U.S. College Tryout," *Christian Science Monitor*, December 27, 1957, 6.

34 J. K. McKay, *Guardian* (London), July 3, 1961, 15.

35 Colomina, *Privacy and Publicity*, 160.

36 "Hyperbolic Paraboloids of Lumber Are Analyzed," *Progressive Architecture* 41, no. 5 (May 1960): 115.

37 "The House That States a New Tradition in Wood," *Living for Young Homemakers* (February 1961): 43–72.

38 Sprague, "Beauty, Versatility, Practicality," 175.

39 "Spectacular Roof Design," *Progressive Architecture* 37, no. 4 (April 1956): 1.

40 "Where Tomorrow Begins," *Progressive Architecture* 38, no. 1 (January 1957): 10.

41 "A Great Architect's Memorial," *Life*, September 22, 1961, 82.

42 Matt Novak, "Recapping 'The Jetsons': Episode 06—The Good Little Scouts," *Smithsonian Magazine*, October 29, 2012, https://www.smithsonianmag.com/history/recapping-the-jetsons-episode-06-the-good-little-scouts-99204721/.

43 Tyler Sprague, "Eero Saarinen, Eduardo Catalano and the Influence of Matthew Nowicki: A Challenge to Form and Function," *Nexus Network Journal* 12 (2010): 249–258.

44 Rob Whitehead, "Saarinen's Shells: The Evolution of Engineering Influence," in *Proceedings of the IASS-SLTE 2014 Symposium "Shells, Membranes and Spatial Structures: Footprints," 15–19 September 2014*, Brasilia, Brazil, ed. Reyolando M. L. R. F. Brasil and Ruy M. O. Pauletti (2014).

45 "Here and There and This and That," *Pencil Points* 5, no. 1 (January 1924): 71.

46 "Mail Order Education: A Look Inside the Thriving Correspondence Schools," *Changing Times* 4, no. 2 (February 1950): 32.

47 "Mail Order Education," 32.

48 Mario Salvador, "Thin Shells," *Architectural Record* 116, no. 1 (July 1954): 174.

49 Seymour Howard, "Useful Curves and Curved Surfaces 1–3," Time-Saver Standards, *Architectural Record* 118 no. 2 (August 1955): 209.

50 Seymour Howard, "Useful Curves and Curved Surfaces 37, 38," Time-Saver Standards, *Architectural Record* 125, no. 2 (February 1959): 245–249.

51 Howard, "Useful Curves and Curved Surfaces 37, 38," 245.

52 Howard, "Useful Curves and Curved Surfaces 1–3," 209.

53 "Hung Roofs," *Progressive Architecture* 37, no. 3 (March 1956): 106.

54 Edward X. Tuttle Jr., "Hypar gambits," *Progressive Architecture* 52, no. 3 (March 1971): 89–91.

55 Richard Stepler, "Hypar Structures: Light, Easy to Build, Cheap—and Permanent," *Popular Science* 216, no. 2 (February 1980): 74–77, 168.

56 Stewart Brand, *Whole Earth Catalog* (Menlo Park, CA: Portola Institute, 1968).

57 Steve Baer, *Dome Cookbook* (Corrales, NM: Lama Foundation, 1968).

58 Brand, *Whole Earth Catalog*, 15.

59 Ant Farm, *Inflatocookbook*, 2nd ed. (printed by the author, 1973), 14.

60 Michelle Kempson, "'I Sometimes Wonder Whether I'm an Outsider': Negotiating Belonging in Zine Subculture," *Sociology* 49, no. 6 (2015): 1081–1095.

Chapter 10

1 Quoted in Sarah Williams Goldhagen, *Louis Kahn's Situated Modernism* (New Haven: Yale University Press, 2001), 72.

2 Ludwik Fleck, "The Problem of Epistemology," in *Cognition and Fact: Materials on Ludwik Fleck*, ed. Robert S. Cohen and Thomas Schnelle (London: Springer, 2012), 81.

3 John McHale, "Global Ecology: Toward a Planetary Society," *Zodiac* 19 (1969): 174–179.

4 René-Just Haüy, *Traité de minéralogie* (Paris: Chez Louis, 1801).

5 Jean-Nicolas-Louis Durand, *Précis of the Lectures on Architecture*, trans. David Britt (Los Angeles: Getty, 2000), 78.

6 Jacques Guillerme, "Notes pour l'histoire de la régularité," in *L'art du projet: Histoire, technique, architecture* (Warve, Belgium: Éditions Mardaga, 2008), 238 (my translation).

7 Guillerme, "Notes pour l'histoire de la régularité," 234.

8 Otto Lehmann, *Flüssige Krystalle* (Leipzig: Verlag von Wilhelm Engelmann, 1904).

9 Spyros Papapetros, *On the Animation of the Inorganic: Art, Architecture, and the Extension of Life* (Chicago: University of Chicago Press, 2012), 121.

10 Esther Leslie, *Liquid Crystals: The Science and Art of Fluid Form* (London: Reaktion Books, 2016), 26. Plateau's work on the subject may be found in his *Statique expérimentale et théorique des liquides soumis aux seules forces moléculaires* (Leipzig: Clemm, 1873).

11 Ernst Haeckel, *Kristallseelen: Studien über das anorganische Leben* (Leipzig: Alfred Kroner Verlag, 1917).

12 Jonathan Massey has compellingly revealed the architect Claude Bragdon's prodigious efforts in the use of the crystal both ornamentally and theoretically. Jonathan Massey, *Crystal and Arabesque: Claude Bragdon, Ornament, and Modern Architecture* (Pittsburgh: University of Pittsburgh Press, 2009).

13 "La nature nous montre parfois la façon dont se construisent ses formes par le jeu réciproque des forces internes et des forces externes." Amédée Ozenfant and Le Corbusier, "Vers le cristal," *L'Esprit Nouveau*, no. 25 (July 1924). Author's translation.

14 László Moholy-Nagy, *The New Vision: Fundamentals of Design, Painting, Sculpture, Architecture* (New York: W. W. Norton, 1938), 122.

15 "Une nouvelle dimension pénètre notre conscience scientifique et plastique." *De Stijl* 7, no. 79–84 (1928): 21–22.

16 From Theo van Doesburg, "Elementarism (The Elements of the New Painting)," *Abstraction-creation: art non figuratif* 1 (1932): 39.

17 Theo van Doesburg, "The End of Art," *De Stijl* 7, no. 73–74 (1926): 29–30.

18 Hermann Weyl, *Symmetry* (Princeton: Princeton University Press, 1952), i.

19 György Darvas, *Symmetry: Cultural-Historical and Ontological Aspects of Science-Arts Relations* (Basel: Birkhäuser, 2007), vii.

20 Lionel March and Philip Steadman, *The Geometry of Environment: An Introduction to Spatial Organization in Design* (London: RIBA, 1971), 345. Robin Evans, "Mies van der Rohe's Paradoxical Symmetries," *AA Files* 19 (Spring 1990): 67.

21 These materials were reassembled for a 2008 exhibition at London's Wellcome Collection; review in Colin Martin, "Exhibition: Design Crystallized in the 1950s," *Nature* 452, no. 815 (April 17, 2008): 815.

22 Maria Bottero, "Questo numero," *Zodiac* 19 (1969): 5; English translation, 223.

23 Anne Tyng, "Geometric Extensions of Consciousness," *Zodiac* 19 (1969): 130–161.

24 Tyng, "Geometric Extensions of Consciousness," 160.

25 Moshe Safdie, *Beyond Habitat* (Cambridge, MA: MIT Press, 1970), 59.

26 Safdie, *Beyond Habitat*, 243.

27 Anne Griswold Tyng, "Proposed College Dormitory," *Zodiac* 19 (1969): 166.

28 Tyng, "Proposed College Dormitory," 166.

29 Tyng, "Proposed College Dormitory," 166.

30 John Rudolph, *Scientists in the Classroom: The Cold War Reconstruction of American Science Education* (New York: Palgrave, 2002).

31 Christopher J. Phillips, *The New Math: A Political History* (Chicago: University of Chicago Press, 2015), 2.

32 Paul Dowling, *The Sociology of Mathematics Education* (London: Falmer, 1998), 170.

33 Phillips, *New Math*, 4.

34 March and Steadman, *Geometry of Environment*, 7.

35 William Huff, "Kahn and Yale," *Journal of Architectural Education* 35, no. 3 (Spring 1982): 23.

36 Goldhagen, *Louis Kahn's Situated Modernism*, 66.

37 Cornelie Leopold, "Precise Experiments: Relations between Mathematics, Philosophy and Design at Ulm School of Design," *Nexus Network Journal* 15, no. 2 (2013): 368.

38 Leopold, "Precise Experiments," 369.

39 William S. Huff, "The Parquet Deformations from the Basic Design Studio of William S. Huff at Carnegie-Mellon University Hochschule fur Gestaltung and State University of New York at Buffalo from 1960 to 1983," manuscript, University of Buffalo Libraries.

40 Harvard University, *Visual and Environmental Studies Catalog, 1976–1977* (Cambridge, MA: Harvard University, 1976).

41 Arthur Loeb, *Space Structures: Their Harmony and Counterpoint* (Boston: Birkhäuser, 1976), xix.

42 Amy Edmondson, *A Fuller Explanation: The Synergetic Geometry of R. Buckminster Fuller* (Boston: Birkhäuser, 1987).

43 Victor Schlegel, *Theorie der homogen zusammengesetzten Raumgebilde* (Dresden: Druck von E. Blochmann & Sohn, 1883).

44 Loeb, *Space Structures*, 114.

45 Peter Pearce had an intellectual lineage which primed him for crystal approaches. He graduated from IIT's Institute of Design, the successor school of Moholy-Nagy's New Bauhaus, in the early 1950s. He was a designer in the Eames Office from 1958 to 1961, during which time he helped develop several major seating designs, including the ubiquitous airport seating system produced by Herman Miller. In 1961 he began teaching at California State University–Northridge, and by 1969 he was at USC, then a nexus for design science due to the work of Ralph Knowles's Natural Forces Laboratory and Konrad Wachsmann's ongoing investigation of universal assembly systems.

46 Peter Pearce, *Structure in Nature Is a Strategy for Design* (Cambridge, MA: MIT Press, 1978), xiv.

47 Susan Pearce and Peter Pearce, *Experiments in Form: A Foundation Course in Three-Dimensional Design* (New York: Van Nostrand Reinhold, 1980).

48 Huff contributed three articles: "Symmetry: Man's Aesthetic Response—Man's Contemplation on Himself," *Oppositions* 3 (May 1974); "Symmetry: Man's Conceptualization of the Universe," *Oppositions* 6 (Fall 1976); and "Symmetry: Man's Observation of the Natural Environment," *Oppositions* 10 (Fall 1977).

49 Keith Critchlow, *Order in Space: A Design Sourcebook* (London: Thames and Hudson, 1969), 3.

50 Z. S. Makowski, "Space Structures and the Electronic Computer," *AD: Architectural Design* 36, no. 1–6 (January 1966): 8–9.

51 Tomás Maldonado, "Tomás Maldonado," in *Ulm Design: The Morality of Objects*, ed. Herbert Lindinger (Cambridge, MA: MIT Press, 1991), 222.

52 William Huff, "The Computer and Programmed Design: A Potential Tool for Teaching," in *Design and Planning 2: Computers in Design and Communication*, vol. 2, ed. Martin Krampen and Peter Seitz (New York: Hastings House, 1967), 102.

53 Ludwig Rase, "Ludwig Rase," in *t-5 Tendencies 5 Tendencije 5: Constructive Visual Research, Computer Visual Research, Conceptual Art* (Zagreb: n.p., June 1, 1973), 89.

54 Pearce's Synestructics had a clear affinity to Buckminster Fuller's spaceframe company, Synergetics.

55 Peter Pearce, interview by author, Los Angeles, August 2, 2013.

56 Pearce, interview by author.

57 Alan H. Schoen, *Infinite Periodic Minimal Surfaces without Self-Intersections* (Washington, DC: NASA, 1970).

58 Safdie, *Beyond Habitat*, 49.

59 Janos Baracs, introduction to *Topologie Structurale* 1, no. 1 (1979): 8.

60 Fleck, "The Problem of Epistemology," 101.

61 Eisenman was selected for this project in response to a request for proposals from Carnegie Mellon University in January 1988, and the state of Pennsylvania was slated to provide an initial $17 million grant for the project.

62 Cited in Peter Eisenman, "Eisenman Architects," A+U: *Architecture and Urbanism* 220, no. 1 (January 1989): 9–52.

63 W. Daniel Hillis, *The Connection Machine* (Cambridge, MA: MIT Press, 1985).

64 Peter Eisenman, "Carnegie Mellon Research Institute," *A+U* 220 (January 1989): 34.

65 Roald Hoffmann, "Fluorite," *The Sciences* 28, no. 5 (September–October 1988): 31.

66 Robin Evans, *The Projective Cast: Architecture and Its Three Geometries* (1995; repr., Cambridge, MA: MIT Press, 2000).

67 Evans, *Projective Cast*, 353.

68 Steve Baer, *The Dome Cookbook* (Corrales, NM: Llama Foundation, 1968), 1.

69 Greg Lynn, *Animate Form* (Princeton, NJ: Princeton Architectural Press, 1999), 16.

Chapter 11

1 A. O. Scott, *Better Living through Criticism* (New York: Penguin, 2016), 149.

2 Derek T. Whiteside, "Wren the Mathematician," *Notes and Records of the Royal Society of London* 15 (July 1960): 107–111.

3 Quoted in René Spitz, *Hfg Ulm: The View behind the Foreground: The Political History of the Ulm School of Design, 1953–1968* (Berlin: Edition Axel Menges, 2002).

4 Editorial in *Progressive Architecture* (July 1971): 55.

5 Antoine Pevsner, "Science Foils Poetry," *Leonardo* 10, no. 4 (Autumn 1977): 324.

6 Ada Louise Huxtable, "Architecture Criticism," *Proceedings of the American Philosophical Society* 134, no. 4 (December 1990): 461.

7 Jose Ortega y Gasset, "Man the Technician," in *History as a System* (New York: W. W. Norton, 1941), 111.

8 Clarice Lispector, *Agua Viva*, trans. Stefan Tobler (New York: New Directions, 2012), 15.

Acknowledgments

The development of this research has been a delightful journey, and I have been distinctly fortunate in my interlocutors and fellow travelers along the way. Foremost among them was my editor Cynthia Davidson, whose incisive and fresh editorial eye as well as patience during the gestation period of the text gave the project the proper room and encouragement. Sanford Kwinter's initial faith in the project was a catalyst without which it would not have come to fruition.

In developing the text, Esther Choi and Alexander Porter offered rigorous and literate insights that helped to smooth rough edges and to focus and deepen critical lines of argument. Several colleagues generously read portions of the text, providing challenging and invaluable feedback that helped me to raise the ambition of the project. Among them Preston Scott Cohen, Roberto Botazzi, Penelope Haralambidou, Bryony Roberts, Michael Osman, Andrew Goodhouse, Iman Fayyad, Cameron Wu, George Legendre, Ed Eigen, Antoine Picon, Matthew Allen, and Greg Lynn were particularly generous.

Interviews or correspondence with Peter Pearce, William Huff, Erwin Hauer, and many others proved critical to unravelling the hidden relationships and affinities that bordered on folklore. Students in my course Narratives of Design Science, in which I refined some of the arguments, were stimulating interlocutors and insightful respondents to many of the arguments presented here. Tobias Nolte, my partner in the design office Certain Measures, was a constant companion in the working out of implications of these ideas in the context of design practice.

The research would not have been possible without institutional support from The Graham Foundation, The Canadian Centre for Architecture, The MacDowell Colony, and the Harvard University Graduate School of Design. The libraries and archivists of Ulm HfG, Institut Henri Poincaré, Science Museum London, Cambridge University, and Houghton Library at Harvard were particularly helpful in the researching of primary resources. I would specifically thank Valerie Bennett from the Architectural Association archives, with whom my wife Michelle and I spend notably productive and enjoyable hours among the thesis archives and Robin Evans materials. Claire Djang and Gia Jung were indispensable and meticulous in the tracing and securing of the image rights from all corners of the globe.

Though they have now been almost entirely rewritten, portions of the text or related ideas appeared previously in the articles "A Machine Epistemology in Architecture," *Candide: The Journal of Architectural Knowledge*, no. 03 (2010); "Design Hacking: The Machinery of Visual Combinatorics," *Log*, no. 23 (2011); "Landscapes, Spaces, Meshes," in *Architecture Is All Over*, edited by Esther Choi and Marrikka Trotter (New York: Columbia Books on Architecture and the City, 2017); "Grayboxing," *Log*, no. 43 (2018); "Form Logics," in *(Non-)Essential Knowledge for (New) Architecture*, edited by David L. Hays, 306090 no. 15 (New York: 306090, 2013); and "The Machinic Animal 1970," in *When Is the Digital in Architecture?*, edited by Andrew Goodhouse (New York: Sternberg Press, 2017).

Finally, and most importantly, I thank my wife Michelle Lee, whose patience and good humor went beyond rational limits. They were unmerited but unforgettable kindnesses.

Index

Page numbers in *italics* refer to illustrations.

Albers, Josef, 235, 237, 363
  Structural Constellations, 237–239,
    238, 241–242
  structural sculpture, 242
Alberti, Leon Battista, 37–38, 58
  De pictura, 143
  De re aedificatoria, 37–38
  linear perspective, 16, 124, 143, 190
algorithms, 23, 25, 241, 345, 370–371
allometry, 268–269
Ant Farm, 339–340
Archimedean solids, 161, 355
architect, 37–38
Architectural Association (AA),
  London, 264, 359
Architectural Record, 24, 245, 312, 328, 340
  geometric tutorials, 318, 331–335
  "Time-Saver Standards," 333, 335
artificial intelligence, 25, 380, 390
atlas, 21–24, 174, 390
  stereoscopic, 150, 156
Auerbach, Felix, 21, 167–168, 172, 185

Babbage, Charles, 55
Baer, Steve, 14, 345, 355, 376
  Dome Cookbook, 339–340, 376, 383
Banfield, Arthur Clive, 172
  photoratiograph, 168–169, *170*, *171*
Baravalle, Hermann von, 363
Baroni, Giorgio, 317–318
barycentric coordinates, 124
Bauhaus, 186, 188, 190, 192, 237, 370
Bemis, Albert Farwell, 278–284,
  288, 292, 295
  cubic module, 280–283, 290, 291
  The Evolving House, *279*, 280–283,
    282
Benham, Charles, 167–168, *169*
Bernal, John Desmond, 251,
  260–261, 264
  "Architecture and Science," 261
Bill, Max, 18, 89, 231, 260, 388–389
  Kontinuität, 231

"The Mathematical Approach in
  Contemporary Art," 11–12, 96
"Pevsner, Vantongerloo, Bill," 96
Rhythmus und Raum, 231
  at Ulm HfG, 96, 251
Bion, Nicolas, 40, *111*
black box, 19, 25, 56, 57, 58, 295, 390
Black Mountain College, 234–235,
  237, 241
Bon, Ranko, 269
Bonnier de la Mosson, Joseph, 32
  cabinet de mécanique et de physique,
    32–33, *34–35*
Bos, Caroline, 226, 241
Bourbaki, Nicholas, 98
Bragdon, Claude, 133–134
Bragg, William H. and William L., 156
Brand, Stewart
  Whole Earth Catalog, 338, 339
Breuer, Marcel, 323–324
Brill, Alexander, 71, 92
  geometric models, 71–72, *73*, 78,
    88, 322
Brunelleschi, Filippo, 37, 38

cabinets, 20–21, 23, 26–27, 58
  cabinets de curiosités, 32–33, 80
  manuals vs., 20–21
  Wunderkammern, 25, 32, 58, 349
Callaghan, Jane, 206, *207*, 209
Candela, Félix, 57, 318, 323, 336
Carnegie Mellon Research Institute
  (CMRI), 379–382
Cartesian space, 106, 124, 127, 275, 354
Cartography, 114, 118
Cassini, Jean-Dominique, 115
Catalano, Eduardo, 319, 322–323
  Catalano House, 319–322, 329
catalogs, 12, 17, 20–23, 47–48, 85–88,
  234, 245, 333, 335
  of crystal form, 346
  Cubic Constructions Compendium, 291
  of curtain wall types, 295
  of floor plans, 266

421

Fuller's World Game as, 303
Funktionentafeln, 79–80, *82*, 88
Infinite Periodic Minimal Surfaces without Self-Intersections, 175, 245
International Encyclopedia of the Unity of Science, 187
Le Corbusier's collection of, 88, 328
of mathematical models, 63, 72, 78–84
La méthode graphique dans les sciences expérimentales, 201
Whole Earth Catalog, 338
Center for Cubic Construction (CCC), 288–292, 294, 295, 299
choice architecture, 298
chronophotography, 150–151, 194, 196, 201, 219
clampylograph, 45
Clement, Joseph, 53–55
cloud chamber, 183, 215
Cohn-Vossen, Stephan
Geometry and the Imagination, 85–87, *86*, 234, 354
Colomina, Beatriz, 310, 328
computer, 137–138, 294, 296, 300, 373–385, 389–390
bombsight, 213–214
"The Computer and Programmed Design," 370–371
computer numerically controlled (CNC) fabrication methods, 258
Flatwriter as, 296, *297*
"Impossible Objects," 270
as mathematical encapsulation, 57
perception and, 143, 174–175, 270
as repository of design knowledge, 24–25
stereoscopic drawing and, 174–175
as superintelligence, 303–304
World Game, 303–304
See also artificial intelligence
conchoidograph, 36, 40, 44, 53
conic section, 33, 39–41, 45, 55, 64
conoid, 76, 88, 152, 313, 316, 317, 324
containerization, 16, 288
correspondence courses, 332
Crary, Jonathan, 17, 143
Critchlow, Keith
Order in Space, 369–370
Crum Brown, Alexander, 227
"On a Case of Interlacing Surfaces," 227–229, *230*, 239, 245

crystals and crystallography, 14, 15, 23, 153, 345–351, 369–385
Eisenman and, 379–381
Fuller's airplane hangar, 344
Haüy and, 349, 350
Huff and, 362–363, 369–370
Lehmann and, 351–352
liquid, 351–352, 365
Loeb and, 355, 365–369
Pearce and, 368–369, 374–376
stereoscopy and, 156, 175
thought collective, 374–379, 383–384
Tyng and, 357–359
Ulm HfG and, 363
Van Doesburg and, 353
"Vers le cristal," 252
cubes and cubic lattices/matrices, 275–278
Bemis and *The Evolving House*, 278–284, *279*, *282*
Center for Cubic Construction, 287–291
crystallographic, 156
Friedman and, 295–300, *297*
Gropius's illustration of a projected cube, 190–192, *191*
hypercubes, 28, 133, 353, 382
March and Steadman and, 292–295, *293*
Martin and, 284–287
n-cube, 379–382
Necker cube, 153, 169, 192, 237
voxel, 295, 305
cycloid, 41, 333, 388

Dehn, Max, 234–235, 254
de L'Orme, Philibert, 23, 39
descriptive geometry, 11, 226, 333
courses on, 161, 316
and crystallography, 21
hyperdimensionality and, 126–127
models of, 65–66, 78
Monge and, 46
and stereoscopy, 74, 144, 176
design knowledge, 53, 190, 371
computer as repository for, 24
encapsulation of, 41, 53, 56–58
geometric knowledge and, 47, 51
instrumental knowledge vs., 36, 38–39, 46
De Stijl, 94, 96, 135, 291
De Stijl, 134–135, 353

Dieste, Eladio, 318–319
dilettante, 388–389
Dirichlet domains. *See* Voronoi diagram
DNA, 357, 359
domes
    Dome Cookbook, 339–340, 376, 383
    geodesic, 57, 121, 333
    Santa Maria del Fiore, Florence, 37
drawing, 12, 16–17
    catalogs of, 21
    hyperdimensional, 126–127
    See also light drawing; machines:
        drawing; stereoscopy and
        stereoscopic drawing
Duchamp, Marcel, 131, 133, 161
Durand, Jean-Nicolas-Louis, 64,
    349–351

Earth
    circumference of, 106, 115
    irregular form of, 105, 118, 121, 126
economics
    mathematics and, 284, 287
    modularity and, 277–278
    World Game and, 303
Edmondson, Amy, 368
Eisenman, Peter, 89, 384
    CMRI project, 379–382
ellipse, 39–40, 88
ellipsograph, 36, 41–44, 52, 53
Emde, Fritz
    Funktionentafeln, 79–80, 82, 88
encapsulation, 17, 57–58
    encoding vs., 19–20
    knowledge and, 22, 25, 27, 36–37
    machinic, 37, 41, 46, 48
encoding, 17, 19–20, 58, 104, 390
entasis, 44, 45
equations
    Diophantine, 294
    parametric, 201, 333
Erlangen Program, 21, 23, 85
Ernst, Max, 80–84, 87, 88
    La fable de la souris de Milo, 80–84,
        81, 98
Escher, M. C., 223, 374
Esprit Nouveau, L', 134–135, 329
Euler, Leonhard, 65, 234, 316
Evans, Robin, 354
    The Projective Cast, 14–15, 49,
        103, 324, 383
expertise, 388, 391

Farey, John, Jr., 41–44
Festival of Britain, 172, 354–355
Fleck, Ludwik, 26, 347, 379, 384
formulations, 12
Fouquiau, Paul, 260, 266
Frazer, John, 359
    AA project, *361*
Friedman, Yona, 296–300, 304
frieze groups, 355
Fuller, Buckminster, 121, 333, 339, 354,
        357, 368
    Black Mountain College, 234
    Dymaxion house, 280
    geodesic dome, *122*, 339
    geometric methods, 57, 104
    Synergetics, 374, 376
    triangulation, 121
    World Game, 300–304
functions, multivariable, 103–104, 134,
        135, 136, 137

Gabo, Naum, 89–96, 261
    Linear Constructions, *93*
    Martin and, 94
    mathematical models and, 92
    Project for Bijenkorf, *95*
Gaudí, Antoni, 74, 76
Gauss, Carl Friedrich, 118
geodesy, 14, 15, 118, 125, 127
    See also domes: geodesic
geometry
    geometrical lathe, 51
    geometrical pen, 40–41, 44,
        52, 168, 212
    Geometry and the Imagination, 85–87,
        86, 234, 354
    non-Euclidean, 63, 71, 132, 135,
        187, 315–316
    Platonic solids, 64, 127, 355
    See also descriptive geometry
gestalt psychology, 259, 270
    Albers and, 239, 241–242
    Wertheimer and, 235–237, *236*
Giedion, Sigfried, 123
Gilbreth, Frank and Lillian,
        204–206, *205*
Gould, Stephen Jay, 269
Graatsma, William, 288–291
graphic method, 183, 185, 187–189,
        201–204, 213, 215
grid, 275–277, 281, 283–284
    Cartesian, 106, 124

Center for Cubic Construction and, 288–305
"The Grid as a Generator," 287
world-triangulation, 121
See also cubes and cubic lattices/matrices
Gropius, Walter, 167, 186, 188–192, 280
at the Bauhaus, 186, 192
Moholy-Nagy eulogy, 219
optical illusions, 190–192, *191*, 219, 223
Package House System, 283
Groth, Paul, 21, 156
Groupe de Recherche Topologie Structurale, 376–379, 390
group theory, 345
guilloché, 44, 212

Habermas, Jürgen, 20
Haeckel, Ernst, 21, 352
harmonograph, 167–168
Harvard University
Graduate School of Design, 94, 137, 269, 287, 365
Laboratory for Computer Graphics and Spatial Analysis, 137, 269
Philomorphs, 268–269
Hauer, Erwin, 175, 242–245, 247, 254, 376
Continua Design #1, 243
patents, *244*
Haüy, René-Just, 292, 349–350
Heidersberger, Heinrich, 214–215, *216*, 217
Heinlein, Robert
"And He Built a Crooked House," *10*, 11
helicograph, 42, 44, 53
Henry, Desmond Paul, 214, 215
hexadecimal code, 294
Hilbert, David
Geometry and the Imagination, 85–87, *86*, 234, 354
Hillis, Daniel, 380
Hochschule für Gestaltung (HfG), Ulm. *See* Ulm, Hochschule für Gestaltung
Holmes, Oliver Wendell, Sr., 145–146
homomorphism, 345, 351
housing, 278, 283
Howard, Seymour
"Useful Curves and Curved Surfaces," 332–335, *334*

Hruban, Konrad, 317
Huff, William, 14, 355, 362–365, 383
"The Computer and Programmed Design," 370–371
parquet transforms, *364*
Symmetry, 369
at Ulm HfG, 363
Huffman, D. A.
"Impossible Objects as Nonsense Sentences," 270
hyperbolic geometry, 15, 57, 64, 234, 257, 310–340, 369
Baroni and, 317
Candela and, 57, 318, 323
Catalano and, 319–323, 329
hyperbolic paraboloid (hypar), 24, 313, 315, 317–318, 322, 324, 329–330, 333, 335–336, 340
hyperboloid of revolution, 152, 196, 313, 315–316
Le Corbusier and, 324, *325*
Phillips Pavilion, 88
Shukhov and, 315–317
See also ruled surfaces
hyperdimensionality, 14, 15, 28, 71, 105, 123–127, 131–133, 134–138
De Stijl and, 353
Duchamp on, 131, 133
Eisenman and, 380–381
Gropius and, 189
Harvard's Laboratory for Computer Graphics and Spatial Analysis, 137–138
Jouffret and, 126–127
Lagrange and, 124
Le Corbusier, 132
Poincaré and, 135
Riemann and, 104
Shukhov and, 316

illusion, 27, 153, 223, 245, 259
Gropius and, 189–192, *191*, 219
light drawings, 183, 196, 212, 213
stereoscopy and, 143–144, 146, 161, 179
impossible objects, 192, 223, 242, 270
Inflatocookbook, 339–340
invariants
allometric, 268–269
topological, 224, 234, 266

Jahnke, Eugen, 79–80, *82*, 88
Johnson, Carroll K.
   ORTEP, 174
Jouffret, Esprit, 125–127, 133

Kahn, Louis, 354, 357, 359, 362–363
Kandinsky, Wassily, 133, 237
Kepes, György, 219
Kiesler, Frederick, 357
Klee, Paul, 133, 237
Klein, Felix, 23, 92, 154, 224, 357
   Erlangen Program, 21, 85
   Klein bottle, 69, 169, 224
   mathematical models, 69–72
knots, 132, 226–230, 234, 235,
   239–241, 247
   Arnhem Seifert diagram, *225*, 226
   Seifert diagram, 239–241, *240*
   Tait's "First Seven Orders of
     Knottiness," *228*
Knowledge, 16, 24, 27, 46–47,
   56–58, 98, 346
   diffusion, 158, 328, 331, 335–339
   encapsulation, 17, 19, 22, 33, 40, 52
   instrumental, 36–38, 46, 56, 58
   models and, 72
   perception and, 239
   unity of, 186–188, 197
   See also design knowledge; manuals;
     thought collective
Krasil'nikov, Nikolai, 136–137

Laffaille, Bernard, 316–317
Lagrange, Joseph-Louis, 124, 125
lattices, 345, 353, 355, 368, 370–371
   crystal, 85, 156, *157*, 241, 284, 349
   cubic, 275, 294, 298
   knot, 229, 230
   models, *348*
   surface, 245
Laue, Max von
   Stereoskopbilder von Kristallgittern, 156
Le Corbusier, 13, 87–89, 97, 323, 328
   "After Cubism—Rejection of the
     Fourth Dimension," 132
   Church of Saint-Pierre, Firminy, 88,
     90, *91*, 324
   L'Esprit Nouveau, 134
   Notre-Dame du Haut,
     Ronchamp, 324, *325*
   Phillips Pavilion, 88
   "Vers le cristal," 352

Lehmann, Otto, 351–352
Le Ricolais, Robert, 261, 264, 345, 354
light drawing, 183–184, 219
   analytic and generative, 200
   Heidersberger's rhythmograph, 215
   Lissajous curves and, 201, *203*, *204*
   Marey and, 201–204
   Moholy-Nagy and, 194–200
   motion studies and, 204–212
   See also graphic method
Lissajous, Jules Antoine, 201, 204, 212
Lobachevsky, Nikolai, 315–316
Loeb, Arthur, 269, 355, 363, 366, 374, 383
   Harvard Graduate School of
     Design, 365
   *Space Structures: Their Harmony and
     Counterpoint*, 368
Lynn, Greg, 385

machines, 17, 18–19, 46, 201, 258, 298
   Alberti vs. Brunelleschi, 37–38
   drawing, 33–37, 40–45, 48, 51–55, 58
   seeing, 19, 214, 270
   See also light drawing; stereoscopy
     and stereoscopic drawing
Maldonado, Tomás, 251, 254, 260, 363,
   370, 388–389
manuals, 22, 23–24, 58, 183, 335, 370
   cabinets vs., 20
   Dome Cookbook, 339
   Funktionentafeln, 79–80, *82*, 88
   geometric, 47–48, 310, 333, 338–340
   Géométrie descriptive, 46
   Inflatocookbook, 339
   *Traité élémentaire de géométrie à quatre
     dimensions*, 126
March, Lionel, 14, 87, 89, 284, 292–295,
   354, 376
   graph networks, 266, 269–270
Marey, Étienne-Jules, 150–152,
   201–204, 215
Martin, Leslie, 94, 97, 261, 284–287,
   292, 295
   "The Grid as a Generator," 287
mass media, 16, 24, 310–312, 322, 328,
   331, 335–340
Maxwell, James Clerk, 53, 227
   stereograms, 154, *155*
McHale, John, 347, 357
McLaren, Norman, 169–174, 177
   Around Is Around, 172, *173*
Meinong, Alexius, 223, 259, 270, 271

mesh, 103–104, 106–107, 114–115, 119, 121, 124, 137–138, 194
Meydenbauer, Albrecht, 119
Mies van der Rohe, Ludwig, 294
minimal surfaces, 241, 322, 333, 376
  Neovius and, 70
  Pearce and, 369
  Plateau and, 229, 232, 233, 247, 352
  Schoen and, 175, 245, *246*
  Schwarz and, 275, *276*
Mises, Richard von, 156
Möbius, August Ferdinand, 124, 224, 231
models, mathematical, 11, 16–20, 23–24, 62, 63–64, 97–99
  cabinets and, 32–33
  in *Geometry and the Imagination*, 85–87, *86*
  hyperbodies and, 127, *128*, *129*
  molecular, 357
  stereophotographs of, 156, 229
  See also catalogs; plaster models; string models; topology
modularity, 278, 290, 310, 349, 379
Moholy-Nagy, László, 184, 188, 190, 200, 215, 219
  at the Bauhaus, 192
  crystal, 352
  light drawings, 200, 215
  Light-Space Modulator, 196, *198–199*
  New Bauhaus, *193*, 196–197
  The New Vision, 194
  Things to Come, *195*, 196
Monge, Gaspard, 125, 126, 324
  École Polytechnique, 65
  géométrie descriptive, 46–47, 316
  models, 66, 74, 76, 194
  stereoscopy and, 144, 152, 161
  surfaces reglées, 65–66, 313, 316
Moore, Henry, 242
Morris, Charles, 187, 197, 200
motion studies, 194, 215, 219, 363
  Gilbreths and, 204–206, *205*
  Marey and, 201–204
  NASA and, 209, *210*, *211*
  Shannon and, 206, *208*
Musmeci, Sergio, 247
  Ponte sul Basento, *248*, *249*, *250*

NASA, 175, 209, 376
  MASTIF, 209, *210*
nature, 14, 27, 32, 302, 352, 353, 388, 391–392

Necker, Louis Albert, 153
  Necker cube, 153, 169, 177, 192, 237
Neovius, Edvard, *70*
networks (graphs), 302–303, 349
  meshes and, 103, 121, 138
  topology and, 227, 254, 259–264, 266–269
Neurath, Otto, 186–187, 197
New Bauhaus, 196–197
New Math, 16, 362, 384, 389
Nowicki, Maciej, 319–321, 331, 335
  J. S. Dorton Arena (Paraboleum), 319–321
Nuffield Foundation
  hospital studies, 264–266, 268

Olivier, Théodore, 66–68, 71, 74, 76, 78, 88, 92, 152, 194, 315
Otto, Frei, 18, 247
Ozenfant, Amédée, 132, 134, 352

Palmer, Catherine, 206, 209
Paris, 71, 114, 125
  cabinets de curiosités, 32
  École Polytechnique, 65
  Haussmannization, 259
  Institut Poincaré, 45, 96
parquet deformations, 363–365, 374, 382
patents, 245, 317–318, 322–323
Pearce, Peter, 14, 175, 345, 355, 368–369, 373–376, 383
Pei, I. M.
  Luce Memorial Chapel, 324
Penrose, Francis C., 44
Penrose, Lionel and Roger
  "Impossible Objects," 270
perspective, 110, 115, 156, 176
  illusion and, 223
  linear/Albertian, 16–17, 39, 105, 124, 190–192
  stereoscopy and, 143–144, 146
  triangulation and multiplication of, 107, 110
Pevsner, Antoine, 89, 94, 96, 132, 389
photography
  photogrammetry, 119
  photoratiograph, 168–169, *170*, *171*
  stereophotography, 143, 146, 152, 172, 177, 204
  See also stereoscopy and stereoscopic drawing

physics, 16, 17, 21, 24, 26, 53, 63, 357, 392
  crystallography and, 346, 352, 369
  emergence and development
    of, 184–185
  Gabo and, 92
  Gilbreths and, 204
  graphic methods of, 183–
    185, 212–215
  hyperdimensionality and, 133, 134
  light drawings, 200
  Physik in graphischen Darstellungen,
    167–168, 172, 185
  stereoscopy and, 158
  topology and, 227
Picard, Jean-Félix, 114
planimeter, 36, 52–53
plaster models, 21, 63–64, 69, 78, 97–98
  Catalano and, 322
  Klein and, 71–72
Plateau, Joseph, 229, 247
Poincaré, Henri, 118, 125–126,
  127, 134–135
Progressive Architecture, 24, 245, 268, 312,
  328, 329, 331, 335
psychology, 189, 192, 346, 352
  See also gestalt psychology

Ramsden, Jesse, 107
  See also theodolite
Rase, Ludwig, 371–373
Resch, Ron, 257–258
Riemann, Bernhard, 104, 125, 134
Rigge, William F., 213
Rittel, Horst, 266
robots, 206, 380
Rowe, Colin
  "The Mathematics of the Ideal
    Villa," 13–14
Rule, John, 161–166
  stereodrawing device, *142, 164, 164*
ruled surfaces, 55, 94, 194, 206, 312–313,
  315–319, 323–324, 328
  architectural use of, 76–78, 331
  Douliot and, 47–48, 56
  Euler and, 65
  Gaudí and, 74
  Le Corbusier and, 88–89
  Marey and, 152
  models, 64, *67, 68*
  Monge and, 64–69
  as "New Sensualism," 310
  Shukhov and, 315–316

  See also hyperbolic geometry
Russell, Bertrand, 187, 189, 223
  Principia Mathematica, 186–187

Saarinen, Eero, 323, 324, 326–327,
  330–331
Safdie, Moshe, 357, 359, 376
Schoen, Alan, 175, 245, 376
Schwarz, Hermann Amandus, 275, *276*
Seifert, Herbert, 239–241
Shannon, Claude, 206–209
Shukhov, Vladimir, 315–317
Skiagraphy, 78
Slothouber, Jan, 288–291
Snellius, Willebrord Snel van
  Royen, 106–107
space, structure of, 125, 365
spaceframes, 362, 370, 371, 373, 384
stereoscopy and stereoscopic drawing,
  143–179, 229
  drawing, 143, 145–146, 152–179
  in geometric education, 160–161
  invention/origins of, 143–146
  library, 145
  in science, 150, 158
  stereoscopic animation, 169,
    172–174
  *Stereoscopic Photographs of Crystal
    Models*, 156
Storr, John
  Forest Products Pavilion, Oregon
    Centennial, 329
string models, 64, 97, 194, 242, 324
  Gabo and, 92
  Gaudí and, 76
  Klein and, 71
  Le Corbusier and, 88–89
  Olivier and, 66, 69, 315
  See also ruled surfaces
Stubbins, Hugh, 323, 335
Suardi, Giambattista, 40–41, 52,
  168, 212, 214
surfaces
  developable, 65, 94
  Seifert, 226, 239–242
  See also minimal surfaces;
    ruled surfaces
surveying, 53, 103–105, 110–121,
  125–126, 138
  instruments, 19, 33, 106–107, 118, 147
Sutherland, Ivan, 175–177

Tait, Peter Guthrie, 227–229, 235, 239, 266
Tange, Kenzo
    Cathedral of Saint Mary, Tokyo, 324
    Yoyogi Olympic Gymnasium, 323
theodolite, *102, 107, 108–109,* 110, 115, 118, 119
Thompson, D'Arcy Wentworth, 214
thought collective, 26, 347, 376, 379, 383–384, 389
Topologie Structurale, 376–380, 382, 384
topology, 14, 15, 23, 64, 96, 125, 192, 223, 322
    of form, 227–259
    formal vs. functional, 226–227, 251
    of function, 259–271
    Groupe de Recherche Topologie Structurale, 376–379, 390
    models, *252, 253*
    perception and, 192
    Ulm HfG and, 251–257, *252, 253,* 260, 266, 269, 270, 363, 371
Torroja, Eduardo, 317, 318
triangulation, 103–122, 138
    Fuller and, 121, *122*
    Great Trigonometrical Survey of India, 115
    Meydenbauer and, 119
    in military fortification/Vauban's city walls, 110–114
    origins of, 106
    surveying, 115–118
Tyng, Anne, 14, 18, 345, 357–360, 363
    City Tower project, *357, 358*
    dormitory project, *360*
    molecular geometry and, 359, 373

Ulm, Hochschule für Gestaltung (Ulm HfG), 26, 89, 363, 370, 388
    Bill and, 96
    Huff and, 363
    Maldonado and, 251, 260
    topology and, 251–257, *252, 253,* 260, 266, 269, 270
unity of science, 186–187, 189, 197

Van Berkel, Ben, 226, 241
Van Doesburg, Theo, 134–136, 242, 291, 353–355, 384
Vauban, Sébastien Le Prestre de, 110–114, 121
vaults, 39, 41, 48, 313, 317, 319, 330

Vienna circle, 186
virtual reality, 143, 175–177
virtual volume, 150, 168–169, 194, 196, 200, 201, 204, 209, 213–215
virtuoso, 388–390
vision, 94, 104–105, 107, 137, 206, 237–241, 392
    Gropius and, 189–194, 200
    hyperdimensionality and, 124, 126
    machine, 25, 270, 390
    stereoscopy and, 143–144, 147, 153, 166, 176, 179
    surveying and, 110, 114, 118
    training of, 65
    Wertheimer and, 235
    See also gestalt psychology; illusion
volute, 45
Voronoi diagram, 260, 368

wallpaper groups, 355
wargames, 277, 302
Waste in Industry, 278
Wertheimer, Max, 235
Weyl, Hermann, 354, 369
    Symmetry, 239, 354, 369
Wheatstone, Charles, 144, 146, 153, 179
Whitehead, Alfred North, 58, 89, 187, 189
    Principia Mathematica, 186–187
Wilson, Charles Thomson Rees, 183, 185
Wren, Christopher, 313, 388
Wright, Frank Lloyd, 78, 280

Xenakis, Iannis, 88

Zeischegg, Walter, 254–257